odenese

gl'interiori, e così caldo piglia le buo
acqua, tienli in luogo caldo, che non g
tro dì, poi piglia la gola, la lingua, e
tto mettilo in un tegame, e dagli di ta
piglia tutto il magro della carne de
el coltello scarna e separa il magro d
caio il magro, che hai cavato, e pesa le
ob 7 di sale, e mezza di pepe la me
codella di vino rosso, e impasta il tut

PIGS AND PORK

COPYRIGHT © 1996 BIBLIOTHECA CULINARIA S.R.L., LODI

EDITOR: DANIELA GARAVINI
GRAPHICS AND DESIGN: LORENA TORTORA, ARCENCIEL NEW
PHOTOGRAPHY: NICOLETTA INNOCENTI
COVER PHOTOGRAPH: AGENZIA LAURA RONCHI, TONY STONE

COPYRIGHT © 1999 FOR THE ENGLISH-LANGUAGE EDITION
KÖNEMANN VERLAGSGESELLSCHAFT MBH
BONNER STR. 126, D-50968 COLOGNE

ENGLISH TRANSLATION: ISABEL VAREA FOR ROS SCHWARTZ TRANSLATIONS, LONDON
EDITING/TYPESETTING: GRAPEVINE PUBLISHING SERVICES, LONDON
PRODUCTION MANAGER: DETLEV SCHAPER
ASSISTANT: NICOLA LEURS
PRINTING AND BINDING: KOSSUTH PRINTING HOUSE CO.
PRINTED IN HUNGARY
ISBN 3-8290-1463-5
10 9 8 7 6 5 4 3 2 1

PIGS AND PORK

90 RECIPES FROM ITALY'S MOST CELEBRATED CHEFS

Preface by Fausto Cantarelli
Introduction by Alberto Capatti
Text by Daniela Garavini
Wines selected by Giuseppe Vaccarini

KÖNEMANN

THE PIG THROUGH THE AGES

Fausto Cantarelli

In ancient Roman times, the valley of the River Po in northern Italy was famous for its *silvae glandiferae*, the oak forests considered to be the ideal terrain for large-scale pig farming. The Po Valley not only produced meat for the local population and for the capital, but also supplied the army. Earlier still, in the Neolithic Age, domesticated pigs grazed beneath the shady trees of the *terramare*, the prehistoric lake settlements characteristic of this part of Italy, excavated in the late 19th century by the archaeologists Strobel and Pigorini. In those days the herds grazed under the watchful eye of the *porcarius*, the swineherd. According to the edict published in 643 AD by King Rothari of Lombardy, the *porcarius* was the most highly valued of those whose job it was to look after animals. He knew that only fatty meat would keep and retain its tenderness and flavor. This is why it became customary to lay in stocks of meat, so that it could be eaten at any time or carried on long journeys.

Pigs were naturally greedy and the acorns and beech nuts they found in the woods soon fattened them up. Even today, in the Spanish sierras, the animals still feed in this way. In Italy, however, the custom of putting pigs out to pasture soon died out, so marking the end of the first age of the pig. Because the woods had been reduced to mere coppices, they no longer provided enough to sustain the animals. At the same time, because of the increase in population, there was less food than in earlier centuries.

So began a new age in which, to make pig farming a more profitable business, pigs were reared in small, dark sheds

 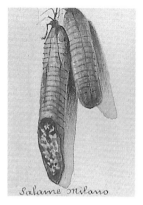 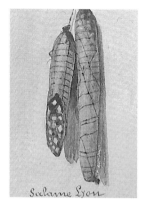  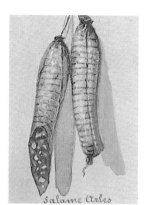

Cotichini   Salame Milano   Salame Lyon   Salamini   Salame Arles

A selection of salamis, c. 1900. Watercolors from the A. Bertarelli Public Print Collection, Milan.

adjacent to the house. Later they were consigned to sties next to the diary.

The efforts of poor farmers and their enthusiasm for making food production more efficient meant that their simple cuisine was able to compete with that of the ruling classes, which had been dominated by oriental spices since the days of the great Roman gastronome Apicius. Meanwhile the humble countryside dwellers were developing more and more regional products and gradually building Italy's rich and varied culinary traditions. All this was possible because the pig had proved to be an adaptable creature, so much so that farmers devised ways of supplementing the more expensive meat from the back of the animal with the less sought-after front cuts to create *prosciutto*.

According to Charles Darwin, who disagreed with his contemporaries, the pig was a descendent of the European boar and the Asiatic hog, which the British used to improve the local breeds. In the 19th century, under the patronage of Francis I of Bourbon, local Pelatella pigs from the Neapolitan region were crossed with Yorkshires, the first crossbreeds intended to produced larger porkers.

Our modern era is the third age of the pig. In a global market driven by economic forces, variety is increasingly being eroded, and there is a trend toward a depressing uniformity in what we eat. If this continues, it will mean the decline not only of the pig, but also of Italy's rich gastronomic heritage.

Problems began to arise when it became clear that the market could not keep pace with consumer demand. The responsibility lay with those who were prepared to forego quality for the sake of convenience. Cooked sausages and salamis are likely to destroy demand for the finest matured meat. If this trend continues, the outcome could be disastrous, so it is important to maintain an appreciation of quality products.

Basically, human beings have always lived happily alongside the pig, through self-interest rather than altruism, making as much use of the animal as possible.

I believe that this book will be welcomed by those who still need convincing that their attitude to pigs and pork products has been mistaken and damaging, and who now wish to establish new rules more suited to the competitive global market, dedicated to enhancing the pleasure of eating. Alberto Capatti's persuasive introduction permits the reader to rediscover the traditions surrounding the relationship between people and pigs. Delving into folk literature, the author has collected some fascinating anec-dotes about the traditions of rural life. He also passes on time-honored ways of preparing the meat of this supremely available, tasty, and friendly animal.

Every bit of the pig can be eaten, from the tail to the trotters. All it takes is a little time and a few simple ingredients to bring out the deep and delicious flavors and create a veritable feast, in a culture partial to a snack of warm pork scratchings and all the other heavenly gifts bestowed by the generous carcass of the pig.

Cappelletti

Coppa

Polpettone

Mortadella

THE EPITOME OF SENSUAL PLEASURE

Alberto Capatti

Few domestic animals have given rise to such a wealth of associations as the pig. We think of it as omnivorous and at the same time good to eat. No other animal has been the subject of so much literary moralizing. Even today, the word "pig"

evokes images of sexual debauchery and gluttonous hedonists lounging at the table in a state of semi-undress, gorging themselves on mounds of food with absolutely no qualms. Appealing, appetizing and above all allusive, the pig speaks to our senses, our stomachs, our tastebuds and our imagination. Destined from birth to be the victim of traditional cultures, the pig has always been sacrificed to human greed, to fulfil our desire for meat. It is also the butt of caustic humor. One example of this is the recipe, devised by the Futurist painter Fillìa, for *porcoecciatato* – loosely translatable as "excited pig": A raw salami, with the skin removed, served up on a dish of boiling hot espresso coffee mixed with lashings of Eau de Cologne.[1]

One of the virtues of the Futurists was that they never attempted to reconcile artistic expression with the dictates of good taste. Their facetious literary efforts are liberally peppered with phallic symbolism; less frequently, the state of arousal is emphasized by stimulants (coffee) and by the use of alcohol and essential oils (Eau de Cologne), as in Fillìa's recipe.

However, the Futurists were right to imbue the pig with the qualities of modern man i.e. excitable, ludicrous and impulsive. But country- and city-dwellers alike have always taken a more utilitarian attitude towards the pig, an animal which redeems itself by being so very useful. Fat and dirty, certainly, it is also plump and appetizing. Until the birth of microbiology, pig meat was regarded as splendidly sumptuous fare, enjoyed not only by the poor but by the great and the good as well. In medieval Latin literature the comic *Testamentum porci* ("pig's testament"), celebrated the redemption of this messy creature. Many benefited from the pig's death: the departed animal willed that his flesh should be seasoned with the salt of his own wit, his snout provide a hearty meal, and his dung left to the peasants. He also invited mourners to assuage their grief by enjoying his carcass, nicely spiced with pepper and honey.[2] In another version, his bristles were left to the Jews, who always treated him with respect, his whiskers to painters to make brushes, his bones to gamblers to make dice, and his trotters to gardeners to enrich the soil. In the Reggio Emilia region of Italy, the *Testament d'un nimel* (testament of a pig) is still a prizewinning theme in traditional and popular poetry.[3]

A list of goods and chattels was not the only way of charting porcine history. At different times and in different places, sayings and customs have grown up around the pig, proclaiming the pros and cons of pork. First came the popular etymological association between pigs and dirt, *porcus quasi sporcus*,[4] repeated ad nauseam for centuries, then exemplary tales illustrated everything from the origin of the species to its ritual significance. Today, it might seem ridiculous to inquire where and when pigs first appeared on earth. Nevertheless, this is the type of question our insatiable quest for knowledge has always raised. To find the answer, it is worth consulting the *Breve catologo delle cose che si mangano e bevono nuovamente, ritrovaro* (brief catalog of food and drink, rediscovered), written in 1548 by Ortensio Lando:

"Laberius of Egypt was the first to eat pig, whose flesh Galen preferred above all others; if it is castrated it grows fatter. It cannot live in Arabia, it is dedicated to Ceres, and at gatherings of friends it was customary to slaughter a pig."[5] There is much truth in this humourous quote about the love of Greek doctors for such fatty meat and the instinctive antipathy of the Semitic peoples toward the animal's dirty habits.

As for the Egyptian homeland of the first pork-eater, this suggests that the pig goes back to earliest Antiquity and, according to Lando, the Egyptians were the first to consume the pig's ancestor, the boar. He also mentions fattening and preparing the pig for sumptuous feasts with friends and family. This was a whole ritual which spawned numerous legends about the pig's willingness to sacrifice itself and allow every scrap of it carcass to be gnawed and sucked. In his book *On Agriculture*, Varro, the great Latin scholar, commented "It is said that pigs have been given by nature for feasting",[7] or in Vincenzo Tanara's enthusiastic and somewhat free translation "The pig was given to us to add joy to eating."[8]

In 1644, a century after Lando and 1,700 years after Varro, Tanara, an agronomist from Bologna, published *L'economia del cittadino in villa* (The economy of the city-dweller) and it seems that all three writers speak with one voice and use the same language to describe the succulent subject of the pig.

Not to be outdone by his forebears, Tanara refers to 50 ways suggested by Pliny and Plutarch of "seasoning this precious meat". Having quoted several of these, he then goes on to add 110 recommendations of his own. Beginning with a recipe for 15-day-old suckling pig roasted on a spit, the main ingredient then increases in age and weight, culminating with a full-grown porker, which the author spices with philosophical and archaeological comments. The oldest and most famous method of preparation was *Porco troiano*, a whole pig, stuffed with meat like the legendary Trojan horse, which hungry Romans attacked with forks and teeth. "The pig's stomach was crammed with fig-peckers, their entrails, eggs and heads and other varieties of costly meat."

Traditional ways of seasoning, cooking and serving are not always so far removed from modern ones. Some of the more unusual methods include festive rustic dishes like "half a boiled pig's head" which "will make a family of poor men fat". For those who preferred something even simpler, there was suggestion number 76: "bread fried in lard until crisp", a mainstay of the urban population. Tanara rambles on endlessly about the many virtues of the pig, its sociability, generosity and general excellence, and how it can be prepared according to the financial means of the consumer. In a world where fat was an object of desire, sellers of pork fat, grocers and pork butchers did a roaring trade and could swagger around the place basking in admiration, however much they might be suspected of dealing in stale sausages and rancid headcheese. Vincenzo Tanara records the fate of animals aged between six months and two years, roasted whole in the oven and stuffed with fragrant herbs. "In the provinces of Umbria and the Marches, every Sunday morning in the town and village squares, roast meat was sold, which was of great benefit to the poor, for without needing to use a cooking pot they could buy a small piece to savor with their families."[9]

Apart from selling meat on the street, the butchers' knowledge of anatomy and skilful cutting meant that they could supply enough to satisfy the appetite – and suit the purse – of everyone from the king to his swineherd. Every bit of skin, meat and offal had its own flavor, its special devotees, and its own price. The cheapest part of the carcass was the head, followed in ascending order by the brain, the collar, the neck, the tongue, the liver, the ribs, the breast, the shoulder, the loin, the belly, the *prosciutto*, and, best of all, the blood, which was transformed into black pudding and blood sausage. These parts were then subdivided into even more pieces, except for the nerves and cartilage, which were no use to anyone. By cooking and seasoning with salt and spices, the butchers managed to produce ever more sophisticated delicacies, like various kinds of mortadella and tasty *cervellata* (made with pork and pig's brain). These Milanese specialties were a quintessential mix of strong flavors, including lean meat, spices, sugar, grated Parmesan or Lodi cheese, a discreet

hint of saffron and endless other exotic ingredients, such as pine nuts and raisins.

While the art of sausage-making consisted of cramming and sealing a blend of flavors into a skin, roast pork had the honor of being cooked over fire and complemented by fruit. To provide contrasting flavor, prunes, sour cherries, sliced apples and pears, pitted olives, and grape must were used to stuff suckling pigs, or to add a touch of excitement to stews. The pig supplied fat and meat, the cook supplied fruits and fragrance.

There were as many mouth-watering manifestations of pork, "the epitome of sensual pleasure", as there were regions and traditional regional products. Every poet and essayist was well aware of this culinary scenario. For the first banquet described in the first book of *Baldus*,[10] Teofilo Folengo prepares *prosciutto* from the Abruzzo, headcheese from Naples, *offelle* from Milan, and French sausages. Tommasi Garzoni is equally geographically aware in Discourse CXXII of *La Piazza universale di tutte le professioni del mondo*, a review of trades and professions in which he rattles off a list of itinerant vendors bacon fat vendors: grocers, sausage sellers and poulterers, and

an array of products: spicy Lucanian sausages, Milanese *cervellate*, musky Trevigiano sausages, salami from Piacenza, mortadella from Cremona, and "new" sausages from Modena, different from those from Bologna.[11]

A good deal of space is devoted to the Emilia region, because even in the 14th century, there was a bewildering selection of black mortadella from Modena, served hot, and sausages made in Bologna "in the style of Modena". Having exhausted the products of the Po Valley, the connoisseur then sets off on a tour of Tuscany, beginning with typical Florentine salami, made with lean meat and seasoned with garlic. The mouth-watering smell of sausages leads in yet more directions, beyond the numerous Neapolitan varieties of belly of pork, toward the ancient tradition of the Lucanian sausage.

While the place of origin was a guarantee of quality, consumption of pork also depended on the calendar. "The domestic pork season begins in November and continues throughout carnival",[13] but this was in fact just the beginning, the coldest time of year. Peasants ate pork infrequently, saving it for special occasions, usually during those brief periods known in Romagna as "*le nozze del porco*" (the pig's wedding). Writing in the 18th century, Gerolamo Cirelli reported that the peasants "celebrated with as much abundance as their poverty would allow, and sometimes on such occasions consumed more than half a pig, for they crave this food more than any other."[14] Pork fat and lard were eked out for as long as possible. Fresh pork fat was salted, desalted, sliced, pounded, rendered over the fire, or boiled or melted to make lard which would keep from the fall until well into the summer, turning increasingly yellow. It had all sorts of culinary uses. Professional cooks had their own favorite methods for making the most of its flavor. One cook would grind it with cabbage, another would desalt the fat in wine and crush it with herbs to make soup, while a third would melt it in broth to create a tasty brew known as *lardiero*. The lifetime of the product did not depend only on the date when

the pig was slaughtered and conditions under which it was stored but also on the customs dictating when people could indulge their taste for pork fat. It was not actually true to say that everything finished at Carnival time. In Christian countries, the length of time lard could be kept was subject to a double religious rule. The first season lasted until Shrove Tuesday and the second began after the 40th day of Lent.

Butchers, sausage-makers, and friers all waited eagerly for Christ's ascent to Heaven. Olivier de Serres recalls how, on the day before Easter, a pork fat market was held in front of Notre-Dame in Paris, where all the master pork butchers had stalls.[15] The season lasted until the end of spring, allowing the French to satisfy their desire for the first, largest and firmest peas, dressed with pork fat. Another fate awaited fresh pork. Far fewer animals were slaughtered in the heat of summer and, after the French Revolution of 1789, the practice was prohibited by law. Not until the early 20th century and the advent of refrigeration could pork be sold during the hot summer months.

Although geography and time of year were prime factors in pork consumption, the flavor and choice of the accompanying ingredients varied enormously. The history of food is dominated by the pig. Let us look at an example from French history: the etching that illustrates the first issue of *L'Almanach des gourmands*, published in 1804, shows a library, whose shelves are lined, not with books, but with foodstuffs such as *prosciutto*, suckling pigs, and large *andouilles*, garnished with truffles. The implication is that products such as these speak volumes about the rules of excellence and so are preserved intact, bound in skin, so that they can be consulted in order to gain knowledge – "tasted", so to speak. The author, Grimod de la Reynière, who defines the pig as a "civilized boar", advises that suckling pigs should be of such an age "as to cause of the fair ladies to swoon when it is presented to them at a banquet". Not only does the library provide shelf space for *prosciutto*, the arrival of a piglet has the same effect as Cupid's arrow.

The question of origin, quality and, last but not least, the keeping qualities of pork deserves closer examination. Pork could be considered as the gastronomic equivalent of wine, whose place of origin has to be identified and the right level of maturity ascertained. Experts must determine the right

Cover of La Salameide *by Antonio Frizzi, printed in Venice in 1772.*

moment to consume it and in what quantity, as well as judging the flavor. Where this procedure takes place seasonally, it serves to define an artefact and gives it a life of its own, setting a standard and a benchmark.

Bearing in mind that pork has been the only traditional Italian meat to be uniformly distributed across the land and to supply cooking fat and animal protein for most of the year, its importance in the shaping of our society's tastes is undeniable. We have always been able to recognize and classify wine by the combination of color, smell and taste. The pork sausage, on the other hand, has always had a traditional "bouquet" of salt, spices, and preserved meat. An etching illustrating *La Salameide* by Antonio Frizzi, printed in Venice in 1772,[17] shows a gentleman sampling a freshly cut salami. He holds it in his left hand and raises it to his nostrils, adopting the traditional pose of the wine-taster.

The study of pork is a fertile source of information on the nutritional qualities of the meat and its associated products. Lard and pork fat were used in all kinds of dishes by people of every social class, whether in broth, game stew, or meat pies made with pork or other meat; with roasts, trussed and larded

and cooked over the open fire, or in sweet bread rolls served at the beginning and end of the meal. Tanara quotes a time-honored recipe: "Lard blended with flour makes a sugary, honeyed and quite delicious morsel."

Because lard combined so well with practically any other ingredient and the ease with which quantities used could be controlled, it was simple enough to produce all manner of substantial, soft, aromatic things to eat.

Quite apart from this excessive use of fat, skilled cooks had developed the technique of exposing the skin and meat of the

pig to red hot flame until the gelatinous juices emulsified to create the characteristic crispness of suckling pig. Because the consistency and moistness of the meat varied very considerably according to whether it was boiled, smoked or cured, it was possible to devise an amazing array of tasty dishes, from spit-roasted leg to boiled trotter, from meat loaf to galantine. At the same time, pork readily absorbed salt, spices and other seasonings, and was receptive to heat, which stimulated the flowing of the juices, with delicious results. It is tempting to make a connection between this versatile and satisfying effect on the palate and the distinctly porcine bellies and cheeks of people who gorged themselves on such delicacies.

Several hundred years have elapsed since Garzoni wrote about his 16th-century itinerant vendors, or Tanara described his 17th-century town-dwellers. Their outmoded recommendations for the preparation and consumption of pork and its fatty by-products would not be well-received today. Nowadays, lard, pork fat, and their by-products such as *ciccioli* and *sfricoli* – pork fat scraps used to make lard, known as *grasò* in Reggiano dialect – are regarded as indigestible and only to be eaten occasionally. But they were very widely used until the late 17th century, when modern Italian cuisine began to develop. So there is a sense of historical continuity. Here is what Tanara has to say in his recipe number 64: "Fresh pork fat is melted to make lard, with which thousands of foods are made to be relished. Those minuscule pieces which remain after melting and when the fat is boiled, when mixed with pastry, make excellent morsels, which we call *crescente*."

The same recipe appears in the first edition of Pellegrino Artusi's *La scienza in cucina*, published in 1891: "When I heard the word 'crescent', I thought they were talking about the moon: in fact they meant the flat loaf, the *schiacciata*, *focaccia* and fried bread with which we are all familiar. The only difference is that, in Bologna, to make it softer and more digestible, when kneading the flour with iced water, they add a little lard."

It is tempting to hail Artusi as the last heir to the 16th-century tradition of Tuscany and Emilia, but that would be mistaken. All early 20th-century cuisine was influenced by the fact that the traditional local choice of fats and meats was gradually being supplanted by an inter-regional food distribution system and new ideas about diet. The need for some kind of map of food sources and a catalog of the regional use of fat was voiced in the first edition of *La Scienza in cucina*, in an introduction to a history of Italian taste: "For frying, people use the fat which is best produced in their own part of the country. In Tuscany they prefer oil, in Lombardy the preference is for butter, and in Emilia the choice is lard, which is made extremely well there. It is pure white and firm, and the scent is enough to cheer the spirit with a single sniff".

Hence, the unusual practice in that region of frying young pullets in lard, with tomatoes. For frying I prefer lard, because it has a more pleasant, aromatic flavor than oil."[18]

The different use of cooking fats by the middle classes in the central north of Italy illustrates a twofold revolution. Firstly, the gradual decline in the use of pork fat and the lard obtained when it was melted and clarified, and, secondly, the increasing preference for what were regarded as leaner cuts, such as the loin in the case of fresh meat, and *prosciutto crudo* in the case of cured products. Pork underwent a change of image. It was no longer seen as meat which could be cooked in any number of ways, boiled, stewed, fried, oven- or spit-roasted. Instead, the preferred method of preparation was roasting without the addition of any fat. The blade shoulder, a popular joint in the 16th century, was consigned, like loin and spare ribs, to the oven. The Florentine version of roast loin of pork, *àrista*, appeared in Artusi's and Agnetti's books of recipes collected from the now-unified Italy. It was described as a good family dish, which could be eaten cold and would keep in the larder for several days.[19] In Venice, the same joint was marinated in milk, spit-roasted and served hot.[20] Rather than devise new cuts, turn-of the-century butchers would stick to the old favorites, reserving the fat "for browning potatoes and reheating greens".

Contemporary instruction manuals for pig-breeders provide an even more fruitful source of information on pork consumption trends than recipe books. These textbooks start with the premise that "it is not customary to boil pork". Old traditions were revolutionized: *lardiero* broth gave way to lighter consommés; flame was the best heat source for preparing dishes for which the main requirement was that they should be "simple".

Despite the theory that every part of the pig was edible, trotters went into an unstoppable decline and the pieces that had once been boiled ended up being stewed with vegetables. Pig's pluck (the heart, spleen, liver and lights) was considered inferior to calf's. Writing in 1914, Marchi and Pucci declared that pork "is above all roasted" and "good specimens with exquisite fat give off an aromatic smell."[21] Even the more highly-prized and ostentatious cuts were eclipsed. Fewer and fewer tables were graced with salted *presciutto*, which Tanara recommended cooked in wine, and which Grimod de la Reynière sweetened with Malaga before roasting and sprinkled with yet more Malaga before serving whole. With no concordat in force in Italy, religious taboo was replaced by

prejudice on grounds of hygiene. Paolo Mantegazza pronounced pork virtually indigestible and unwholesome.[22] Some accepted his withering condemnation, although with the advent of refrigerated warehouses to store the carcasses, local authorities finally allowed freshly slaughtered meat to be sold in summer. With the demise of the "pork season" came increasingly stringent food regulations and a growing climate of suspicion. Inspectors appeared, with thermometer in hand and a health certificate for each animal. In a debate dominated by concerns about micro-organisms, there were warnings against raw or undercooked pork. Cooks were

advised to observe the color of cooked meat – pictures of pink or red pork were no more than artistic license – and to rediscover the virtues of old-fashioned boiling, pork scratchings and a ham bone. Housewives turned into "health police" in a new campaign that really did no more than revive

and repeat hackneyed clichés. No pork for obese relatives, sedentary husbands or delicate children. Be sure to chew everything slowly and patiently. The best instruments for sterilization are the grill and the grid-iron. As Lidia Morelli counseled in 1935, "it is not unusual for pork to contain the larvae of parasites like trichinella and tapeworms, which can develop in the human body. For this reason it is prudent to eat only thoroughly cooked pork."[23] It was the beginning of a new era, dominated by the specter of the tapeworm, epitomized by tough chops, burnt to a frazzle.

The blood which peasants in the Romagna made into sweet fritters began to disappear from urban cuisine, as did the Tuscan specialty, which wealthier people such as Pellegrino Artusi enhanced with honey, chocolate, candied fruits and almonds to make black pudding.[24] Paradoxically, however, the old fondness for soft pork rind did not dwindle, as we see from a confession in an unexpected place. After lamenting the fall from favor of boiled pork and the decline of the pig's trotter, and applauding the fine, yielding, succulent texture of fresh pork, the author of a manual on pig breeding admitted: "Where I come from, we use lots of pork rind and *prosciutto* bones in bean soup. Whether it is customary or not, I find it quite delicious and prefer it to all the galantine in the world."[25] This is not a schizophrenic refusal to accept anything new, but a plea to pig farmers to save a few pieces of rind. Some sectors of the food preserving industry would fulfil the task of perfecting time-honored products like sausages and adapting them to contemporary tastes in a market where quality and choice were the watchwords. Until the end of

World War I, Italians still regarded large-scale slaughter-houses, like those in Chicago and Berlin, as something out of science fiction. In Italy, traditional family businesses continued to practice their craft, using small machines for mincing meat and stuffing sausages. While few still believed the myth that every scrap of the pig was edible, certain questions remained unresolved. By 1914, people were asking how to determine the best-quality bristle, and what to do with the skins and trotters. Now there was a new movement towards producing pigs lean and refined enough to compete with and triumph over their bovine rivals. It was also necessary to find the correct kind of fodder for such animals.

After 1920, pigs' snouts and tails became rarities and eventually disappeared completely from pork butchers' shop windows. Restaurants no longer served *hure de sanglier*, the famous headcheese made from the head of the wild boar, once regarded as food fit for a king. At the same time, the list of sausages appeared to lengthen. Salami and sausages mark the final chapter in the age-old history of the pig. Antonio Frizzi resorted to verse in 1772 to trace the origins of one such product, *cotechino*: This is a translation of his poem:

Our Cotichin walks, side by side like a brother, with the zampetto of Modena. His shirt is not made out of gut but from the pigs very own skin. The skilful Geminian[26] craftsman takes the pig skin and finely stitches it along the shoulder and breast, then wraps the leather around the trotter to create a bag, which he fills with chopped mixed meat to make a Cotichin.[27]

While they had stopped eating the eyes, genitals and other appendages of the pig's anatomy, people again wanted large joints of meat which really looked as though they were part of an animal. It began with the *zampone* (stuffed pig's trotter). As a result of the outstanding skills of Modenese craftsmen in

sewing pigskin, from the early 1900s the city enjoyed increased prosperity from manufacturing products for a burgeoning sector which demanded attractive, tasty, calorific foods which were easy to prepare and also projected an image of affluence.

This stitched-together imitation of a pig's trotter soon became the provincial petit-bourgeoisie's answer to the great Westphalia and York hams, when these disappeared from the table. With the revival of interest in regional culture, this new trend took off, and the royal chef, Amedeo Pettini, included it, along with *cazzuola di maiale alla milanese*, a rich stew with pig's trotters, in his book of Italian recipes.[28] It featured on restaurant menus, with Madeira sauce and garnished with spinach,[29] and was served for Sunday lunch or as a warming dish, with puréed potatoes or lentils.[30] Leftover *cotechino* sausage or *zampone* were served cold as appetizers, or hot with sauerkraut or potatoes.[31]

Against all the odds, pork was rehabilitated in a new guise. Some may object that *zampone*, boiled or pre-cooked, makes no demands on the skills of the cook. Maybe, but it does serve to show how, in the 20th century, the ingenious use of pigskin revived an ancient tradition.

Any eulogy of the king of the beasts, which is not, as everyone believes, the lion but the pig, usually ends up sounding ridiculous. In other words, it seems somehow false, like a performance by a second-rate actor who cannot decide whether to plump for a traditional or a modern style. We confuse the porcine snout with refined human features. Today's slimmed-down pig, and even *zampone*, still cannot escape this anthropomorphic association. As long as we cook and serve pork, there will always be a certain empathy between ourselves and the pig.

Let us now end this introduction and move on to the rest of the real *testament d'un nimel*, to the legacy of the animal which is so central to the traditions of the Emilia Reggio region. Having learned a little of the historical background of pigs and pork, it is time, with the help of professional chefs, to learn how this succulent meat can be prepared to suit modern tastes.

Notes

1   F.T. Marinetti and Fillìa, *La cucina futurista*, Milano Longanesi 1986, p. 209 (1st ed. Sonzogno, Milan 1931).

2   *Nugae venales*, Lonini sumptibus societatis 1741, p. 39.

3   *Testament d'un nimel* by Luciano Pantaleoni di Corregio came third in the competition held in the province of Reggio Emilia in 1982 for poems in dialect on the subject of regional traditions surrounding the pig. See *Il maiale nella cultura contadina e nella tradizione popolare reggiana*. Published by Ufficio Agricultura del Comune di Reggio Emilia, Felina 1982, p. 60.

4   "Verres, quod grandes habet vires. Porcus, quasi spurcus", *Isidoris Hispanensis episcopi Etymologiarum sive originum*, XII,1,25, Oxonii 1962. The author discusses the same etymology in Italian and Spanish.

5   Ortensio Lando, *Commentario e Catalogo*, Bologna, Pendragon 1994, p. 124.

6   "Domesticorum porcorum carnes secundum Galenum in libro de subtilitiva dicta omnium quidem ciborum sunt nutrivissime," *Incipit compendium de naturis et proprietatibus alim entorum per magistrum Barnabam de Regio*, in: *L'eccellenza e il trionfo del Porco* edited by Emilion Faccioli, Reggio Emilia Mazzotta 1982, p. 20.

7   Marcus Terentius Varro, *On agriculture*, Loebb classical library 1960, p. 161

8   Vincenzo Tanara, *L'economia del cittadino in villa*, Venice 1680, p. 365.

9   Vincenzo Tanara, *ibid.* p. 162.

10 "Persuttos huc terra suos Labruzza recarat, / huc ve suppressadas Napoli gentilis et offas / Millanus croceas et quae salcizza bibones / cogit franzosos crebras vacuare bo tecchias", Teofilo Folengo, *Baldus*, I, Verses 469-72, Torino Einaudi 1989, p. 32.

11 Tommaso Garzoni, *La piazza universale di tutte le professioni del mondo*, Venice 1610, p. 355 (1st ed. 1586).

12 Vincenzo Tanara, *ibid.*, p. 169.

13 C. Evitascandalo, *Libro della scalco*, Rome 1609, in *L'eccellenza e il trionfo del porco*, cit. p. 84

14 Girolamo Cirelli, *Il Villano smascherato*, p. 26 in Giovanni Batarra, *Pratica agraria*, Rimini Ghigi 1975 (reprint of the 1782 Cesena edition).

15 "On the day before Easter a pork fat market is held in Paris in front of Notre-Dame and all the master Pork Butchers of Paris have stalls there. From Chalons in Burgundy large quantities of fine pork fat are brought to Paris, prepared in the Parisian style. From Normandy and Lower Brittany too, but differently prepared...", Oliver de Serres, *Le theatre d'agriculture et mesnage des champs*, Paris 1605, p. 835.

16 "Bibliothèque d'un gourmand du XIXᵉ siècle", *Almanach des gourmands*, Paris Maradan 1804; Grimod de la Reynière, *Manuel des amphitryons*, Paris Métailié 1983, p. 24 (1808).

17 Antonio Frizzi, *La Salameide poemetto giocoso con le note*, Venizia Zerletti 1772 (reprint published by l'Accademia italiano della cucina, Ferrara 1983).

18 Pellegrino Artusi, *La scienza in cucina*, Firenze Landi 1891, p. 100.

19 Pellegrino Artusi includes the recipe for *àrista* in the second edition of 1895 (*La scienza in cucina*, Torino Einaudi 1970, pp. LXXI and 339). The same recipe appears in *Nuova cucina delle specialità regionali* by Vincenzo Agnetti (Milano società editoriale 1909, p121). Both authors refer to the derivation of the name from the Greek *aristos* – excellent.

20 Vincenzo Agnetti, *ibid.*, p. 49.

21 E. Marchi - C. Pucci, *Il maiale*, Milano Hoepli 1914 (3rd edition), p. 402 and 397.

22 Paolo Montegazza, *Piccolo dizionario della cucina*, Brigola, Milan 1882, p. 94; Mantegazza and Neera, *Dizionario di igiene della famiglie*, Bemporad, Florence 1901, p. 260.

23 Lidia Morelli, *Nuovo ricettario domestico*, Milano Hoepli 1935, p. 147.

24 Pellegrino Artusi, *La scienza in cucina*, cit., p. 321. For the use of pig's blood, see Olindo Guerrini, *L'arte di utilizzare gli avanzi della mensa*, Formiggini, Rome 1918, p. 170-173

25 E. Marchi and E. Pucci, *ibid.*, p. 402.

26 Tassoni's word for a Modenese.

27 Antonio Frizzi, *op. cit.*, p. LXIII.

28 Amedeo Pettini, *Manuale di cucina e di pasticceria*, Casale Marescalchi 1914, p. 310.

29 Touring Club Italiano, *Manuale dell'Industria Alberghiera*, Milan 1923, p. 665.

30 *Ricette di Petronilla*, Olivini, Milan 1938, pp. 102-103

31 Olindo Guerrini, *op.cit.*, p. 167.

32 "A more ancient preconception in Egypt crowned the lion king of the beasts. This is unjust, for the king of the beasts is the pig...", *ibid.*, p. 164.

PIGS AND PEOPLE

DANIELA GARAVINI

The Italian word for the pig is *maiale*, by which is meant the animal. Meanwhile, the person who behaves like one is called *porco*. In France, the animal is a *porc* and the uncouth individual a *cochon*. In English, "pig" applies in both the animal and human sense. "Swine" is widely used as an insult, while the meat is "pork". *Schwein* is the German word both for the animal and the meat, and also a term of abuse.

Clearly, the pig has become part of our real lives and the world of our

imagination. It begins in childhood when pigs appear as leading characters in fairytales. Far from portraying them as physically and morally unclean creatures, the stories almost always emphasize their strength and intelligence (we all remember how the Three Little Pigs managed to outwit the Big

Bad Wolf). These two positive attributes, strength and intelligence, occur again, albeit in a negative sense, in another, much more recent literary classic, *Animal Farm*. In choosing pigs to lead the rebellion and then the dictatorship, George Orwell is clearly passing moral judgment on the people whom the animals are intended to represent. Even so, the author makes many perceptive observations about the nature and behavior of pigs.

Zoologists also tell us that the pig is an eminently malleable beast, both

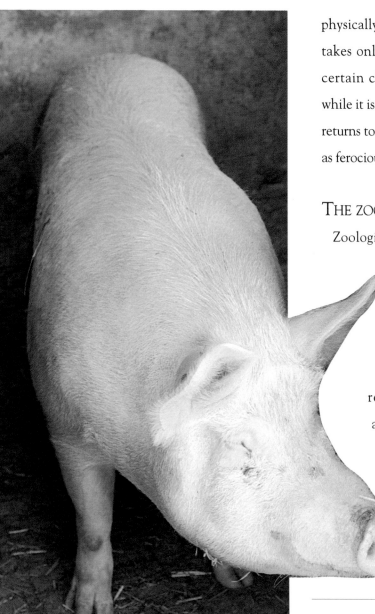

physically and psychologically. First, it takes only a few generations to select certain characteristics and, secondly, while it is docile in captivity, as soon as it returns to the wild it once more becomes as ferocious as its brother the boar.

THE ZOOLOGICAL FAMILY

Zoologists define the pig (*Sus scrofa domesticus*) as an artio-dactyl mammal (one that has an even number of toes) of the Suidae family. The members of the family are recognisable by their long, almost tusklike, canine teeth, and by the position on the abdomen of teats and testicles. They repro-duce by multiple births. Omnivorous

creatures, they have a highly developed sense of smell and hearing, an excellent sense of touch, particularly around the snout, and not very good eyesight. The family also has some wild relatives, such as the African river hog (*Potamochoerus porcus L.*), the warthog (*Potamochoerus aethiopicus*), also native to Africa, the babirusa (*Babirussa L.*), which lives in Indonesia, and the American peccary (*Tayassus pecari L.*). The relationship with the wild boar (*Sus scrofa ferus*) is even closer, and some experts believe that they are the same species. Although they look very different, boars can be crossed with pigs and produce fertile progeny (unlike the mule, or hinny, the result of crossing a male donkey with a female horse, which is born sterile).

In common with their wild cousins, domestic pigs prefer an environment with plenty of water and vegetation. Their ideal habitat is dense woodland with plentiful acorns, beans and other natural foods, and also plenty of shade and water. Pigs cannot bear the heat and, since they have no sweat glands, they cool down by wallowing in water or, better still, mud, which is even more refreshing. By nature they are not dirty animals, but only become so when there is no water or mud available and they are forced to roll in their own excrement to try to ward off the effects of heat.

A VERY ANCIENT RELATIONSHIP

The domestication of the pig goes back a very long way. The earliest evidence of pig-keeping was found on a Neolithic site at Hallan Chemy, in Turkey, dating from 10,000 BC. The number of piglike bones excavated there would seem to

THE NAMES OF THE PIG

Italian = *maiale, porco*

French = *porc, cochon*

English = *pig, swine*

German = *Schwein*

Spanish = *cerdo, cochino, puerco*

Portugueuse = *porco*

Hungarian = *diszñò*

indicate that the inhabitants ate the meat of these animals on a regular basis and must therefore have had the skills needed to catch and tame them.

We know for certain that the ancient Egyptians kept pigs, exploiting them not solely as food. It seems that, after they had sown their crops, they left herds of pigs to graze in the same field. The ancient Egyptians' attitude to pigs was ambivalent. Like many other animals, the pig was sacred to a particular god, but in this case it was the evil deity, Seth. They also believed that people who led sinful lives on earth would be reincarnated as pigs.

For all these reasons, the Egyptians kept their distance not only from the pigs themselves but also from those whose daily task was to look after them.

Ordinary citizens were discouraged from forming close relationships with swineherds, who became virtual outcasts. They were even forbidden to enter the temple, lest they desecrate the hallowed ground.

Meanwhile, other communities, like Jews and Muslims, have never had any doubt at all about their attitude towards the pig. They regard the pig as categorically unclean. The precepts of their respective religions forbid them to breed pigs or eat their meat. While Jewish law dates back to biblical times, several millennia before the birth of Christ, the Islamic prohibition is much more recent. Some, like the anthropologist Marvin Harris, have sought what might be thought of as a material explanation for the Jewish law forbidding the use of pork as food. According to Harris's theory, in the first millennium BC "the physical conditions for pig breeding would have been unfavorable as a consequence of a deliberate process of environmental change" (deforestation and more land given over to growing crops to meet the needs of the increase in population). Far from being a valuable source of food, the pig was a danger to cultivation and "became a creature that was not only useless, but worse than useless; harmful, a curse, a thing that no one should either keep or touch; a pariah." (M. Harris, *Buono da mangiare*, Einaudi, Turin 1990).

In ancient Greece, however, there was no ban on pigs or their keepers. In Homer's *Odyssey* the first person that Odysseus seeks out after returning to Ithaca is none other than Eumæus, the swineherd. Odysseus embraces him, asks him what has been happening at home and, among other news, is told that a good fat pig is cooked for Penelope's suitors every day. Clearly, Odysseus did not consider Eumæus or his pigs unclean. At the same time, a possible derivation of the Italian *maiale* is that the pig is sacred to Maia, one of the Pleiads (the seven daughters of Atlas), lover of Zeus and mother of Hermes, who was later turned into a star by the jealous Hera.

Returning to Italy, pigs seem to have been the main source of meat for the Etruscans, who are believed to have kept pigs in the wild in the woods and forest covering their territory.

The Romans perfected the art of pig breeding although pork was always food for the élite. There were writings on the subjects of rearing pigs and consuming

pork. Varro, the Roman scholar and writer on agriculture, who lived in the 1st century BC, advised pig breeders about feeding. He gave instructions on how and when the herd should graze, not only in order to maintain the pigs in good health but also to ensure that they yielded the tastiest meat. In the 1st century AD, the Roman soldier and farmer Columella described the most suitable woodland in which to rear wild pigs. Then, of course, there were Apicius's celebrated recipes, and Martial's epigrams on the most flavorsome parts of the pig, his own favorite being the womb of the adult sow.

There was such a heavy demand for pork that some had to be imported from Gaul, where there was an abundance of wild pigs in the forests surrounding the many settlements. Nevertheless, pork appeared only on the tables of the favored few.

The Germanic tribes were also very attached to pigs, so much so that warriors who fell in battle were posthumously awarded a pig. Rather than actually breeding the animals, those whom the Romans regarded as barbarians left them to run wild in the dense woods and forests which then covered much of central Europe.

This state of affairs between people and pigs would continue for centuries, especially after the fall of the Roman Empire, when a less refined food culture prevailed.

THE PIG NEXT DOOR

For very many years, pigs could be seen rooting around the houses, feeding on scraps and leftovers (pigs are omnivorous and have to compete with humans for food). They were also herded into the woods and forests which were their principle grazing grounds: a medieval scholar, Massimo Montanari, recalls how in contemporary legal documents the area of a forest was calculated by the number of pigs it could feed. Physically, these animals looked quite different from the modern-day pig: they were smaller and leaner, with a longer snout and a higher back. They were covered with longer bristles, their flanks were narrower, their bones more solid and their tails straight rather than curly. They were much closer to their wild brother, the boar, than the pigs we know today.

For many centuries, certainly in Europe but also elsewhere (in China for example), the pig was the main source of meat. Other animals were bred, but they were usually raised for other purposes and their meat was only of secondary importance. Cattle, for example, were bred primarily as beasts of burden and for their milk, sheep first and foremost for their wool and milk, chickens chiefly for their eggs, and so on. The pre-eminence of the pig was almost certainly due to its adaptability to its rural surroundings, to its great capacity for reproduction and to its proverbial generosity in providing food. How often do we hear the words "no part of the pig is thrown away"?

Most importantly, the meat and fat, if correctly treated,

would keep for long periods, enabling households to build up valuable reserves of food. Most people only ever ate fresh pork in the days immediately after the pig was slaughtered and often they ate only scraps, since the better-quality meat was set aside for making sausages and other preserved products. A large number of traditional recipes for the preparation and dressing of pork still survive, and many of these appear later in this book.

As can be seen from these ancient recipes, whole suckling pigs and leg and loin of pork featured on the menus of the aristocracy. There is no doubt, however, that at least until the first half of this century, most pork consumed took the form of preserved specialties, such as salami and sausages, and raw, cured, boiled or smoked ham. Until that time, "domestic" breeding was widespread in Italy. Between January and February, the coldest months of the year, the *norcio*, the pork butcher who specialized in slaughtering pigs, would do the rounds. The killing of the pig was a cross between a ritual and a festival, but it meant hard work for everyone involved. Often the slaughter coincided with the feast of St Antony Abbot, patron saint of animals in general and pigs in particular. He is not the only saint traditionally associated with pigs.

PIG BREEDING TODAY

These days, "domestic" pig breeding is the exception rather than the rule. Statistics over the past ten years show a constant decrease in the number of farms but an increase in the number of animals. Instead of innumerable farms, each with only a handful of pigs, there are now relatively few breeders raising large herds. As a result, there is much more pork on the market and consumption of salami and sausages, and fresh meat, has increased dramatically.

The transition from small-scale breeding to factory farming has brought about a series of changes. With the introduction of different kinds of feed, the pig itself has changed, as have the places in which it is kept. Slaughtering techniques are different, and so is the nature of the meat itself.

Original Italian breeds of pig (such as the Black Apennine, the Red, the Casertana, Calabrese, Cavallina Lucana) have virtually disappeared.

A few black pigs have been observed in the wild in Calabria, Basilicata and Sicily," writes Graziella Picchi, a researcher with the Italian National Institute of Rural Sociology, in her very interesting study *L'Atlante dei prodotti tipici: i salumi* (F. Angeli, Milan, 1989). But what has taken their place? In the major pig farming regions, breeding is centered mainly on five breeds, chosen around the middle of the last century: the English Large White, the Danish Landrace, the Belgian Piétrain, and Duroc and Hampshire from the United States (R. Roncalli in *Assalzoo, L'alimentazione animale nella storia del uomo*, Edagricole, Bologna 1995). The 1970s saw the emergence of commercial crossbreeds, the result of interbreeding between these five varieties. One of the virtues of the pig is its physical adaptability, so it is feasible, through genetic engineering, to create individual animals with precise physical character-

istics. This is made easier still by the animal's very short biological cycle. The females can mate at the age of nine months and each sow can produce and wean up to 22 piglets each year. Modern pig farmers look for high birth weight of piglets, size of litter and number of

piglets surviving until weaned, uniform size of piglets, daily growth rate, carcasses with a high percentage of lean meat and therefore of high commerical value (Manetti, Tosonotti, *Scienza del maiale*, Edagricole, Bologna 1984). Every coin has its reverse side and, as Graziella

Picchi comments: "this system of enforced reproduction, combined with intensive farming methods, has caused pigs to lose their characteristic resistance and adaptability to their surroundings, making them predisposed to a whole series of diseases which have to be treated with all kinds of medication (*Atlante*, cit.).

No one seems to have considered carefully the risks and repercussions of allowing local breeds to become extinct.

Genetic variety is a fundamental requirement for the survival of a species, and a policy of allowing only painstakingly selected breeds of pigs to exist may, in the medium and long term, prove to be misguided. Not by chance are there organizations devoted to conserving native breeds. These groups have even gone so far as to propose that increasingly rare Italian pigs should be declared part of the national heritage, in order to preserve the genes and chromosomes of the animals, and the native Italian plants on which they feed.

The most noticeable change in the pig's diet is the ever-increasing proportion of cereals, especially maize and soya. This is what might be termed a philosophical change, a move away from using leftovers or secondary products rejected by humans, towards better-quality cereals. This means, however, that pigs and people are competing for food. Besides grain, pig farmers in the Po Valley also nourish their animals with whey produced in the large cheese factories of Emilia and Lombardy. The availability of whey has been the main reason for the increase in pig farming in the Po Valley, especially in the areas of

large-scale cheese manufacturing. The pattern of agriculture in this region has long been based on the food chain, beginning with cows fed on grass and hay, whose milk, cheese, and buttermilk is then fed to pigs, which in turn produce *prosciutto*. Only a very small minority of pigs are still bred in the wild, foraging for acorns and other fruits of the forest.

The physical conditions in which pigs live have also changed. They are no longer housed in the traditional farmyard sties, but in enormous sheds with no adjacent land, recognizable from the pervasive smell along the roads leading across the vast expanse of the Po Valley.

The move from traditional to factory farms has suddenly transformed animal excrement from rich fertilizer into a waste which now pollutes the bed of Italy's most magnificent river and whose

odor pervades the air for miles around. Fortunately, shrewder pig breeders and local authorities are reconsidering the agricultural use of manure, but the problem is a long way from being solved.

There have also been enormous changes in methods of slaughtering. Instead of awaiting the annual arrival of *il norcio*, live animals pass daily along an assembly line, to emerge as half or quarter carcasses. The slaughterhouses are usually part of the same complex as the breeding sheds.

HOW MANY PIGS?

According to 1993 estimates from the United Nations Food and Agriculture Organization, the worldwide pig population totalled 960 million, an increase of around 218 million over the period 1979-81. Of these, 474 million animals were bred in Asia (China, the world's largest pork producer, accounted for 375 million), 243 million in Europe and 128 in North America (95 million in the United States). The remainder were distributed around the rest of the world.

Data from Italy's Central Institute of Statistics, also published in 1993, reveal a total of 8,396,000 pigs bred and 12, 241,000 slaughtered.

As the statistics show, Italy does not produce enough pork to satisfy domestic demand and there is a constant flow of imported animals, which has a negative effect, particularly on pork destined for use in charcuterie and sausages. "It is easy to imagine the problems created when raw material is imported from abroad, especially in terms of the deterioration of meat which, when transformed into charcuterie, needs considerable quantities of additives and preservatives. Not to mention how the quality of the meat degenerates as a result of the stress suffered by the pigs on their journey to Italy." (G. Picchi, *Atlante*, cit).

Most of Italy's 8 million or so pigs are concentrated in the Po Valley, virtually all of them on factory farms. For example, the province of Mantua which, along with Reggio Emilia, holds the record for pig production, breeds 1,200,000 pigs, 15 percent of the national total. The fact that the province's human population numbers 400,000 gives some idea of the scale of this phenomenon!

MEAT OR CHARCUTERIE

Time was when people could only enjoy fresh pork immediately after the pig was slaughtered. For the rest of the year they ate preserved meat, the alternative to killing underweight piglets. Today, however, specific types of pigs are reared to produce fresh pork. They are killed while still young, no more than seven or eight months old and weighing about 220 lb/100 kg, and the meat they yield is lean and tender.

The most popular cuts are the loin, in the form of loin chops, *àrista* (boned loin) cutlets, spit-roast and so on, and leg, neck, shoulder and ribs. The humbler parts of the pig, from the trotter to the snout, from the rind to the tail, were once main ingredients in traditional dishes but are now eaten only on rare occasions, either because prepar-

ation methods are too elaborate and time-consuming or their calorie content is too high.

By contrast, the heavyweight pigs intended for charcuterie and sausages can weigh in at between 400-450 lbs/ 180-200 kg.

They are not only larger but more mature than those earmarked for fresh meat. The difference is comparable to that between young bullocks and the fully grown adult oxen.

Charcuterie and pork sausage manufacturers play a major role in the Italian agricultural and food industries. Apart form large-scale industrial production, artisans in the various regions are kept busy practicing the different crafts linked to the infinite variety of local specialties.

Each region has its own special charcuterie and virtually every part of

Italy has a string of different recipes for very similar products. This is the homeland not only of the world-renowned Parma and San Daniele hams, but also many other fine *prosciutti crudi*. Italy produces thousands of miles of sausages, of many different sizes, composition and type of seasoning. Then there are all kinds of *mortadella*, *pancetta*, rolled or in whole pieces, headcheeses, spicy *cotechini* boiling sausages, *zamponi*, a whole range of salamis, *coppa* (cured neck of pork) and *capicolli*, as well as the more modest *mortadella de fegato*, made with liver, or *coppa di testa*. To give some idea of numbers, in 1989, more than 150 typical local products based on preserved pork were listed.

PORK IN THE DIET

There is a widespread belief that pork is too fatty and difficult to digest. This is absolutely untrue. Lean cuts of pork, like loin or leg, are lower in calories than comparable joints of beef.

As the table opposite reveals, their fat content is lower or the same as that of beef. Lean pork contains the same levels of cholesterol as beef or lamb, and pork contains only a fraction of the poly-unsaturates, and fewer saturates, than beef. It is worth remembering that pork contains a much higher level of vitamin P (which protects against some skin diseases) than any other meat.

Other cuts, such as ribs, are a different matter. So, too, are sausages. However, in these cases, the cooking method can change the composition of the meat and influence the fat content. For example, grilled ribs certainly contain much less fat than if they are baked in the oven or steamed. Before sausages are added to a dish, they should be pricked with a needle and then briefly fried or boiled to remove much of the fat.

Also worth considering are those seldom-used bits of the pig often enjoyed in poorer homes in the days of our grandparents and great-grandparents. I am referring to the rind, trotters, snout, and so on, all of which include a great deal of connective tissue and require long, slow cooking. They are often difficult to digest, not only because of their fat content but because they contain the indigestible protein, purine.

There is no denying that pork fat and lard, which for centuries made the pig so popular, have become demonized in our modern age of plenty, when so much attention is paid to diet.

As always, it is a matter of degree: it is no sin to add a little bit of lard to a savory sauce, or use it to brown the ingredients for a minestrone. It all depends how much you add, how often and, above all, what else is to be eaten at the same meal! Although it is used increasingly less in cooking, lard really is the best fat for frying because it withstands high temperatures. So is it better to demonize lard and then spend all day nibbling bags of potato chips fried in oil of an unknown composition? Or should we allow ourselves the occasional *fritto misto all'italiano*, correctly cooked and drained, the way it is in every traditional regional recipe?

NUTRITIONAL VALUES

Meat	protein	fat	kcal
Lean young beef	21.3	3.1	113
Lean mature beef	20.7	5.1	129
Lean pork steak	18.3	3.0	100
Lean leg of pork	18.7	3.0	102

Values refer to 3½ oz/100g of edible product; source: National Institute for Nutrition 1987.

THE THERAPEUTIC PIG

Pork fat, especially lard, has long been used as medicine. A friend of mine still remembers her grandmother rubbing lard and aromatic herbs (especially thyme) into her chest, throat, and back when she had a cough or a cold. There was usually a myth or folk tale behind this kind of simple remedy in which lard was used, both for its own healing properties and as a base for the medicinal herbs. One of these was the story of St Antony Abbot, patron saint of pigs, who cured shingles, also known as "St Antony's fire", with a lard poultice. Modern medicine also has reason to be grateful to the pig. Since, like us, the pig is omnivorous, it has been used in studies of diseases caused by unhealthy or excessive eating, enabling doctors to discover the link between an overly rich diet and heart and circulatory disorders. But the pig has an even more crucial role to play in the medicine of the future, as a donor of organs for transplant. There is already talk of injecting pigs with human genes for this purpose, which means that our friend the pig will become even more like us.

PORK-BASED PRODUCTS AWARDED THE DOP
(SEPTEMBER 1997)

Coppa Piacentina	(DOP)
Prosciutto di Carpegna	(DOP)
Prosciutto di Modena	(DOP)
Prosciutto di Parma	(DOP)
Prosciutto di San Daniele	(DOP)
Prosciutto Toscano	(DOP)
Prosciutto Veneto-Berico-Euganeo	(DOP)
Salame della Brianza	(DOP)
Salame di Varzi	(DOP)
Salame Piacentino	(DOP)
Valle d'Aosta Jambon de Bosses	(DOP)
Valle d'Aosta Lardo di Arnad	(DOP)

The *denominazione di origine protetta* (DOP) is a seal of authenticity awarded to specific food products by a special commission of the European Union. The commissioners are presented with documents describing the history and traditions of the product and its connections with its country of origin. These documents must be drawn up by a regulatory consortium of producers, whose rules lay down the recipe for the product in question. All the pork-based products listed here have been honored with a DOP. The only variation from the conditions of past awards is that the raw material, in other words the pig, need not necessarily have been bred *in situ*. Producers are allowed to use animals from other regions of Italy. Unlike centuries ago, when the products were first invented using local breeds, standard pigs are now used. However, the recipe remains unchanged, as do the quantities of added ingredients, the specific methods of cutting or chopping meat and, in the case of salami, lard, the conditions under which the product is matured, and other characteristics.

BASIC METHODS
AND COOKING TIMES

COTECHINO SAUSAGE

Soak the *cotechino* in a bowl of cold water for at least 1 hour before cooking. Then take a large needle and make a series of holes along the entire length of the sausage (do not use a fork, otherwise the skin may burst). Wrap in a clean cloth and place in a pan of cold water. Bring to a boil and simmer over a very low heat. A 1 lb/500 g *cotechino* will need about 2 hours from the moment the water comes to a boil.

PORK RIND

Before cooking, singe the rind over a flame to remove any trace of bristle, then parboil and scrape away any remaining hairs. The trotters (*see* below), snout, ears, and any other pieces attached to the rind should be treated in the same way.

FRESH PORK

For health reasons, pork must be thoroughly cooked. However, a little care is needed to avoid the frequent mistake of overcooking, especially when preparing lean cuts such as chops, cutlets, loin or boned loin. Lean pork needs only a little longer than beef but this, of course, does not apply when steaming or stewing fattier pork.

TROTTERS

Before cooking, singe over a flame to remove the bristles, and blanch in boiling water. Then clean very carefully between the toes and scrape the skin. Boil for about 2 hours, then split in half lengthways.

SHINBONE

Best roasted or braised for between 1½ and 2 hours.

ZAMPONE

Soak for at least 12 hours before cooking. Then follow the same procedure as for *cotechino*, taking a large needle and making a series of holes along the entire length. Wrap in a clean cloth, place in a pan of cold water, bring to a boil and simmer over a very low heat for 3-4 hours.

RECIPES FROM THE 16TH TO THE 19TH CENTURIES

ROSELLI

Epulario

PUBLISHED IN VENICE IN 1533

To cook a suckling pig

The pig must be well skinned so that it is white and clean. Cut the pig along its back, remove the entrails and rinse well. Take garlic, cut very small, some good pork fat, a little ground cinnamon, eggs, and pepper and a little saffron. Mix these together and stuff them into the said pig. Firmly sew and bind the pig and put it to cook on the spit. Roast slowly, so the meat and the stuffing is well cooked. Mix a little brine with vinegar and saffron and take two sprigs of rosemary or sage and baste the suckling pig many times.

DOMENICO ROMOLI

La Singolare Dottrina ...

PUBLISHED IN VENICE IN 1560

Black pudding with pig's blood

If you would make delectable black puddings as we do here, they must be thin and not fat, succulent but not greasy,

16th-century wood engraving.

browned but not scorched. But it is necessary, as soon as the pig is dead, to collect the hot blood in a glazed clay pot. So that no bristles might be caught in the blood, lay a cloth over the pot and pour the blood through it. Sprinkle the blood with fine white flour, pepper and salt, and add two egg yolks, which must not be bad. Wash your hands clean and knead the blood with all these things, breaking up all curdled blood and removing any bristles. Leave the pot to rest in a warm place for one hour, then take fresh pork fat and beat it well to make pure lard. Take a good, flat, tin-plated frying pan and place it on the tripod with good charcoal beneath. Add an abundant quantity of lard and when it boils take a small ladle and thinly spread the blood all over the pan, first stirring it well.

Cover the pan at once, for often the blood will swell and spit all around the pan. Allow to brown and cook underneath. If there is too much fat, remove it. Lay some bay leaves on a flat dish, place the black pudding thereon and sprinkle with pepper and white salt. Above all, eat while hot.

BARTOLOMEO SCAPPI
Opera
PUBLISHED IN VENICE IN 1570

To make pies of all manner of cervellati (pig's brain) and fresh sausage, with different materials
Take fine *cervellati* and place them whole in a pastry case with parboiled artichoke hearts, peeled truffles, large muscat grapes and fresh sloes, then cover the pie and cook it. When it is almost cooked, pour a little sour white grape juice and beaten egg yolk through the hole in the pie, and when cooked serve hot. Similar pies may

be made with all manner of sausages which, if you do not wish them whole, may be cut into pieces. In this way there will be room to add oysters taken from their shells.

ANTONIO FRUGOLI
Pratica e Scalcaria intitolata pianta di delicati frutti
PUBLISHED IN ROME IN 1631

Wild pig's pluck in various soups
When it is fresh and the animal is young, the pluck of the wild pig may be cooked in

various soups. First it must be parboiled, cut into small pieces and browned in good lard with finely chopped onions, or marjoram and other finely chopped herbs. It is served hot with pepper and the juice of bitter oranges. When it is browned as described above, it is placed in a pot with meat broth and sour grapes and plenty of spices. It is served hot, sprinkled with pine nuts and raisins.

CARLO NASCIA
Li Quattro Banchetti destinati per le Quattro Stagioni dell'Anno
MANUSCRIPT OF 1684

Brawn
For brawn made from four suckling pigs, clean them well and remove all the bones. Cut the meat from the bellies into long, thick pieces, and season with salt and pepper, nutmeg, cloves, ginger and a little mastic. Then take the pigs, season them well with salt and spices and lay them out along the table. Add to them four pieces of meat from the belly, part fat, part lean, but

these must be more than the thickness of a finger.

Add a little crushed coriander, marjoram, ground pepper and powdered cinnamon and roll them all together. Add some parsley root and put them on to cook. When they are half cooked, add two jugs of vinegar, but let it be strong with the right proportion of salt and, if you wish to garnish the brawn with vinegar, prepare more than one jug. This dish is garnished with mortadella di Cremona, and fine white sugar, and decorated with a sugar figure which represents the God of Love and, at each corner of the dish, finely carved ham, dressed with sugar syrup.

Antonio Latini
Lo Scalco alla Moderna, overo l'arte di ben disporre li Conviti
Published in Naples in 1692

On the culinary virtues of pork
Immediately the pig is slaughtered, the blood must be carefully collected and strained through a cloth; in order that it should not curdle, all impurities it may contain should be removed. To prepare it, add cinnamon, cloves, ginger, nutmeg, fat cut into small pieces, salt, breadcrumbs, and a little pepper to make it piquant. Then place it in a pig's intestine, first washed and cleaned, adding raisins, pine nuts, a little perfumed water, and salt. The said stuffed intestine is then boiled and will have excellent flavor.

Bartolomeo Stefani
L'Arte di ben cucinare et instruire i men periti in questa lodevole professione
Published in Venice in 1716

Large ham cooked in wine
Place a large ham cooked in wine with cloves and cinnamon on a dish with fragrant herbs. Take a knife and with the point make many holes in the ham. Remove some of the meat from these holes and shred it finely, then return it whence it came. Over the edge of the dish build a pergola as high as a man's arm. Take six gilded vases and in them place six copper wires in the form of an arch and thread into those arches sufficient shelled pistachios to fill them. Then stretch strands made of spun sugar to make a vine with sugar leaves, making bunches of grapes in the same manner. Take sugar, drawn into a narrow thread by cooking, and with this thread attach bunches of pistachios in the form of bunches of grapes. On the pergola, place birds

fashioned from said strands of spun sugar, attach them to the pergola by their feet, so that it appears as though they would peck the meat with their beaks.

MASSALIOT
Cuoco Reale e Cittadino
PUBLISHED IN VENICE IN 1751

Ham pie
Take a good ham, remove the skin, or rind, and bad fat. Cut the shank and remove the bone. Then dress with slices of lard, and slices of beef, fine herbs and spices, slices of onion, and a bay leaf. When it is well larded, cook in a charcoal oven, covered tightly with a lid so that no steam escapes, and cook for about 12 or 16 hours, but let the oven not be too hot. When the ham is cooked, leave it to cool in the same pot. Meanwhile, make a thick dough with a little butter or egg, water and flour. Take the dish in which the pie will be served, and line it with pastry with a strong firm edge. Decorate the edge with small flowers, lilies and suchlike. Bake the

pastry in the oven. Take your ham, remove all the fat around it, and place it in the dish with its juices. Arrange the meat in the pastry shell, filling any spaces with slices of beef and fat. Add a little chopped parsley and sprinkle with breadcrumbs, brown with a red-hot baker's shovel and serve cold.

ANTONIO NEBBIA
Il Cuoco Maceratese
PUBLISHED IN VENICE IN 1783

Pig's tongue in various sauces
Having boiled the tongue with salt for some time, you may wish to serve it with sauce *alla provinciale*. First place the boiled tongue in a stewpan with its juices, sliced or cut in two if you so desire, season with sweet spices, and when it is cooked, add the *provinciale* sauce, for which the method may be found in the chapter on sauces, under tongue. It may also be fried with a ham sauce or any other you may wish. The said tongues are good boiled with wine, hay, and onion, and bay leaves,

and sliced, and brought to the table with abundant piquant sauce. Partridge, young peacock, gray partridge and chicken may also be added.

VINCENZO CORRADO
Il Cuoco Galante
PUBLISHED IN NAPLES IN 1786

Pig's trotters in sauce
Pig's trotters, together with the ears of the same animal, must be cooked in water with bay leaves, rind of green lemon, and ground spices. When it has grown cold, remove the meat from the bone and cut it into thin slices. Make a sauce with vinegar, a little sugar, dried cinnamon, cloves, lemon rind, ground pepper, and bay leaves. When it has boiled and taken up the flavors, pour it through a sieve into another pot. Place the trotters and ears in the sauce and boil until it is well blended and thickened. You will need to have a long wooden box, lined with small pieces of pork fat, into which to put this concoction. Leave to grow cold and then

cover with paper and a wooden lid. This meat may be kept for some six months. When served, it should be sliced.

Pork stuffed with eels

Those of refined taste may eat suckling pig stuffed with pieces of eel, first fried in oil with a touch of garlic, chopped fragrant herbs, fennel seeds, bay leaves, salt and pepper. When this is filled it should be roasted on a spit, basted, as is customary, with oil and water. Serve with a sauce of anchovy oil and pistachios.

LA VARENNE

Il Cuoco Francese, ove viene insegnata la maniera di condire ogni sorta di vivande

PUBLISHED IN BASSANO IN 1787

Ham pie after the Turkish style

Prepare a ham and when it is half cooked in water, and boned, lard the lean parts with chopped pork fat seasoned with sweet spices, and these pieces of fat should be the size of a penholder. Then sprinkle the meat with a little sweet spice and a little bruised white pepper, then make a round pastry shell. Connoisseurs may use puff pastry to make the pie, above all in winter, for this is heartier fare for that season. But this dough is more difficult to work, when the crust is made of puff pastry. When the crust is in the dish, lay in it slices of pork

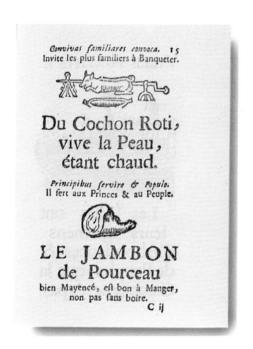

Illustrated primer dating from the late 17th century.

fat, with a little parsley and thyme. Then add the meat and on top of it sprinkle five or six cloves with 12 little pieces of cinnamon bark, and two pinches of ground cinnamon. Add a little parsley, bruised shallots, two ounces of pine nuts and two of raisins, four ounces of shelled pistachios, the zest of a marinated lemon cut into thin shreds, and six ounces of crushed sugar, half a pound of fresh butter, half a pound of lard or half a pound of calf's marrow. On top, lay a large slice of pork fat, a bay leaf or two and a little thyme. Cover the pie, making a hole and a crown on the crust. Cook the pie and observe the sauce from time to time, and add more if need be. Three or four hours before the pie is cooked, you may add mushrooms and blanched sweetbreads and two hours before taking it from the oven, you may infuse a sweet sauce with a glass of white wine, four ounces of sugar, a little powdered cinnamon, and a little vinegar, Note that a pie of this sort may be reheated several times and if it lacks broth, you may add some meat juices.

Il Cuoco piemontese ridotto all'ultimo gusto e perfezione
PUBLISHED IN MILAN IN 1805

Loin of suckling pig in gravy
Cut the loin of a suckling pig into chops and cook in some broth with a bouquet garni, salt and pepper. Take a blanched calf's sweetbread, cut it into large dice and place in a saucepan with mushrooms, some fowl's liver, and some butter. Put all of these things to cook with a good handful of flour dissolved in half broth, half white wine and enough meat juices to give color to the gravy, salt and crushed pepper, a bunch of parsley, small onions, half a clove of garlic, and two cloves. Cook until the sauce is reduced and pour it over the chops. Likewise, you may prepare the chops with gravy and, when they are half cooked, add the calf's sweetbread, the liver and mushrooms with the same seasonings.

FRANCESCO LEONARDI
Apicio Moderno
PUBLISHED IN ROME IN 1807

Pig's ears with pease pudding
Take the pig's ears, place them in an earthenware pot with salt for five or six days, with some bay leaves, thyme, basil, a little juniper, and some ground cloves. Then drain them, wash them, and cook them in water with dried peas. When they are cooked, add a little well boiled and squeezed spinach, pass the purée through a gauze sieve and serve it with the ears; it should be pale green. If you wish to serve them without purée, simply cook them in water and serve with any desired sauce, piquant, Robert, etc.

Galantine of pig's head
Take the head of a young pig, cut close to the shoulder, remove all the bones, remove all the flesh inside and any which remains attached to the rind, and remove the fat. Cut the meat into thin slices and put it on a plate. Season it with salt, fine spices, and powdered fine herbs, some parsley and shallots if you so desire, and a clove of garlic, all chopped small. Now cut two stuffed ox tongues and some raw ham into large slices, and truffles into strips. Take some good green pistachios and peeled sweet almonds. Put the skin of the head in a round saucepan and arrange the aforementioned meats and the fat from the head cut into slices in several layers, and if there is not enough add

some pork fat, slices of ox tongue, ham, truffles, pistachios and almonds.

Continue thus, layer by layer, until the skin is full, seasoning each layer with a little salt and fine spices. If there is not enough meat, add a little pork. Afterward cook the head, shape it into a bag, or roll it like a large mortadella, wrap it in a napkin and tie it firmly with string. Place it in a pot with water, three bottles of boiling white wine, a large bunch of mixed herbs, fragrant ones, four cloves of garlic, six shallots, eight cloves, half a nutmeg, mace, mild pepper, a pinch of coriander, whole peppercorns, cloves, carrots, breadcrumbs, onions, parsley root, two bay leaves and salt. Cook very slowly for about six hours, according to its size. When it is half cooked, place it in a round pot into which the galantine will fit closely, taking the shape of the pot, and place it on a table with a weight of some 15 lbs/7 kg. When it is cold, turn it out and serve on a dish, decorated with a few flowers of your choice and surrounded by a bed of bay leaves and twists of lemon zest. If it is

Instructions on how to carve a suckling pig from Li tre trattati *by Mattia Giegher, 1639.*

wrapped and tied like a mortadella, hang it on a hook with a weight beneath, until it is cold.

VINCENZO AGNOLETTI
Manuale del Cuoco e del Pasticciere di raffinato gusto moderno
PUBLISHED IN PESARO IN 1832

Sausages in the style of Piacenza
Mince 6 lbs/2½ kg of good pork meat with three pounds of fresh pork fat, and season with 4½ oz/120 g of salt, ½ oz/10 g of pepper, two nutmegs, 1½ oz/40 g of cinnamon, 1/16 oz/1 g of cloves and 1/8 oz/2.5 g of coriander, all of them ground. Place the mixture in sausage skins and shape them like small salami which can be dried in the oven or in the air. Some cooks add two crushed cloves of garlic. If they are fresh, these sausages are roasted on a gridiron; if dried, they are boiled in water with wine, a bunch of herbs, and a spiked onion. They may be served hot or cold

and if hot are excellent with sauerkraut or a goodly quantity of greens. This mixture makes excellent sausages, and some in Piacenza add a little ground fennel, or pine nuts and raisins.

Meloncini sausages in the Bolognese style

When you have minced 9 lbs/4.5 kg of pork with 3 lbs/1.5 kg of good fresh pork fat, season the meat with 6 oz/150 g of salt, 2 oz/50 g of whole peppercorns, half an ounce of ginger, ½ oz/10 g of coriander, ⅙ oz/2.5 g of an ounce of cloves, and two nutmegs and fill large sausage skins and bind them in the style of melons and hang to dry. Some cooks use half beef, half pork.

---

cut them into pieces, wash them well and drain away all the water. Cook very slowly in some water. When you think it is cooked, add a good measure of rich, spiced cake, crumbled (cooks call this *smazza-pane*, as well I know!). Take another earthenware dish and put in it some toasted almonds, all chopped, a little belly pork chopped into small pieces, a little shredded lemon zest, a little cinnamon, salt, pepper, and some shelled pine nuts. Add some white vinegar and some sugar to take away some of the sourness. Leave to cook then sample a little, and when it is to your liking add a handful of mint and a little vinegar, and if it is too sharp, a little sugar.

IPPOLITO CAVALCANTI
Cucina Teorico-Pratica
PUBLISHED IN NAPLES IN 1837

Calf's and pig's liver

If you wish, you can make a fine hearty dish of liver. Take the calf's liver, and that of the pig, by which we mean a young one,

FRANCESCO CHAPUSOT
La Cucina sana, economica ed elegante
PUBLISHED IN TURIN IN 1846

Fillet and leg of pork boiled in the style of Campagna

Take 48 oz/1.5 kg of rib or leg of pork and remove 2 oz/50 g of fat from the meat and

chop it up together with an onion, a carrot and a rib of celery, and fry for a moment. As soon as the vegetables take color, add

the pork with 2 pints/3 l of water and salt. Skim, cover the pan well and leave to cook for three hours. Then peel six large potatoes, cut them in half and into pleasing shapes, and cook them for half an hour in another saucepan with a third of the pork broth, then serve them on a platter in a circle around the pork fillets.

This boiled pork is eaten with mustard; and excellent pasta soups can be made with the broth.

GIAMBATISTA BASEGGIO

Celio Apicio delle Vivande e Condimenti ovvero dell'Arte della Cucina

PUBLISHED IN VENICE IN 1852

Roast suckling pig with pasta and apples
Prepare and wash the suckling pig. Mix 1 oz/30 g of pepper with apples and wine and put on to boil.

Break sheets of dried pasta into pieces and throw them into the pot, stirring with a green bay twig until the mixture is soft and thick. Stuff the pig with the mixture; close with toothpicks, wrap in paper and place in the oven, then dress and serve.

Boiled milk-fed suckling pig with cold Apiciana herb sauce
Grind white pepper, coriander seeds, mint and rue in a mortar; moisten with boiled grape must and add apples, wine and seasonings. Pour over the boiled suckling pig while still hot, having first dried it with a clean cloth, and bring to the table.

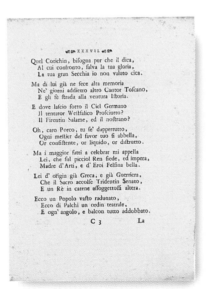

Pages from Gli elogi del porco by Tirgrinto Bistoni, Modena 1761.

GIUSEPPE SORBIATTI
La Gastronomia Moderna
PUBLISHED IN MILAN IN 1855

Pork chops in the rustic style
Prepare eight good cutlets, beat them and put them in a baking pan with a little olive oil and butter, fry for two minutes on each side, and reserve the fat. Pour a glassful of white wine over the meat and cook until reduced, pour over a spoonful of broth, and add salt, pepper and spices. In the fat which you have reserved, fry fine slices of onion and, when they are slightly colored, add them to the cutlets and leave to simmer for three hours. Skim the fat and reduce the sauce to syrup, adding a tablespoonful of birthwort vinegar and a handful of parsley. Arrange the cutlets in a circle and dress with the meat juices.

Large joint of pork on a spit
Take a quarter of a pig thoroughly cleaned of all bristle, and cut it in straight lines, along and across. In a vessel large enough to hold the meat, marinate it in oil, a little

vinegar, salt, two bay leaves, leaving it for a few hours. Then thread it onto a spit that you have the strength to turn, wrap it in sheets of paper lightly greased with oil and tie them firmly with string. Then cook over a moderate flame, basting frequently with oil and a little vinegar, letting it cook for 5 hours. After 4 hours of turning, remove the paper, and sprinkle the meat with breadcrumbs and cook until it is nut brown. Take the meat from the spit and place it on a platter large enough

to hold it. Take the dripping pan, add a spoonful of broth and a very little vinegar, simmer for two minutes, skim off the fat, pour the sauce through a strainer into a sauceboat. The 5 hours indicated is for large joints of pork. Small joints require only 3½ or 4 hours cooking.

Il cuoco milanese e la cucina piemontese
PUBLISHED IN MILAN IN 1859

Grilled pigs' ears
Lightly salt the pigs' ears and cook them in

PPROVA AD UNANIMITA
VOTO DI GUERRA ALLA
ULISSE COLOMBINI DI   PER LO STERMINIO FATT
BOLOGNA •   NELLA NOSTRA CLASSE

good broth, with salt, pepper, coriander, birthwort, fine pork fat and half a glassful of white wine. When they are cooked, allow to grow cold, then cut them in half towards the red part, and rub lightly with their own fat, cover with breadcrumbs on all sides, and brown them on the gridiron.

GIOVANNI VIALARDI
Cucina borghese semplice ed economica
PUBLISHED IN TURIN IN 1863

Sauté of pork with black truffles
Having removed the cartilage and skin from a fillet of pork, cut it into slices two fingers thick, beat it a little and place in a frying pan with butter and brown it all over on a high flame. Add a spoonful of flour and as soon as it browns pour over it half a glassful of Madeira or Malaga wine, and a glassful of broth or water.

Add a little salt, pepper, chopped parsley, the juice of one lemon, and 1 oz/30 g of cleaned and sliced black truffles, or white ones, provided they are good. When the meat is tender and the sauce reduced, serve covered with sauce.

CATERINA PRATO
Manuale di Cucina per principianti e per cuoche già pratiche
PUBLISHED IN GRAZ IN 1892

Pork with horseradish
Meat from the back, breast or shoulder of a young pig, with the rind still attached, is stewed in pieces, with sliced onion, bay leaves, peppercorns and a little vinegar with enough water or broth to cover the meat. At the moment of serving, the reduced broth is poured over the meat, which is then sprinkled with breadcrumbs and grated horseradish.

GIOVANNI NELLI
Il Re di Cuochi trattato di Gastronomia Universale
PUBLISHED IN MILAN IN 1898

Pig's head in gravy
Prepare, rinse and clean a pig's head and divide in two and place in salt in the same manner as ox tongue, except that you should not use saltpeter unless you wish it to be red after 15 or 20 days. On the day on which you wish to serve it, remove the pig's head from the salt and cook it in the same manner as ox tongue. Then remove all the bones and cut into squares or rectangles, leaving aside the excess fat. Serve with a purée of spinach, onions or lentils or whole chickpeas seasoned with reduced sauce *espagnole*, sauerkraut, string beans, or beans *maître d'hôtel*. It is best served with a piquant, remoulade, or tomato sauce, which should be mixed at the moment of serving, with a little sugar, lemon juice and a spoonful of mustard sauce.

55 CHEFS SHARE THEIR SECRETS

There now follow 90 recipes in which pork in its endless forms constitutes the main ingredient. Most of these recipes are for main courses, but there are some for dishes that make ideal starters, as well as recipes for soups, rice and pasta dishes, and even desserts.

Some of the dishes are inspired by traditional Italian regional cooking – a rich source of pork dishes – while others are completely new, either in their combination of ingredients, their method of preparation or their presentation. All are listed according to type (starter, soup or pasta, main course, and so on) in the index.

The 55 contributors are leading chefs working in many different parts of Italy. Accompanying the recipes is expert advice on the best wines to complement each dish, carefully selected by a great *sommelier*, Giuseppe Vaccarini.

ZITI WITH SPICY CALABRESE SAUSAGE

Serves 4

400 g/14 oz of ziti (thick spaghetti), 1 lb 5 oz/600 g of fresh San Marzano tomatoes, peeled and chopped, 4 tbsp of extra virgin olive oil, 1 clove of garlic, chopped, basil and parsley, 2 small, round, green chili peppers, seeded and chopped, ⅓ cup/40 g of pecorino di Crotone cheese, thinly shaved, 3 oz/80 g of nduja (typical Calabrese sausage, made with lots of pork fat combined with leftover pork and strongly but exquisitely seasoned with hot pepper, and stuffed into a thick skin), a very little salt

Method: 1. To prepare the sauce, pour the oil into a shallow frying pan and add the chopped tomato, garlic, chili pepper, basil and just a touch of salt. Cook for about 15 minutes, then add the *nduja*, stir thoroughly and continue to cook for another 3 minutes.
2. Meanwhile, in boiling salted water cook the pasta *al dente*, drain and then stir it into the sauce.
3. Serve sprinkled with pecorino di Crotone and parsley.

Wines: The intensely piquant flavor of this rich, juicy sausage calls for a strong, well-structured, red wine, not too young, very delicate, fairly acidic, with a hint of tannin, and a bouquet of ripe fruits and spices, such as Cagnina di Romagna, Pomino Rosso, or Sant'Anna Isola di Capo Rizzuto Rosso.

PINUCCIO ALIA

OVEN-BAKED PORK WITH HONEY AND PEPPER SAUCE

Serves 4

8 pork loin steaks, 1 cup/250 ml of white wine, 7 tbsp/100 ml of extra virgin olive oil, zest of 1 orange, grated, 4½ tbsp/100 g of honey, 1 tsp of hot paprika powder, 1 handful of wild fennel seeds, 1 red onion, 2 oranges, salt, extra virgin olive oil to dress salad

Method: 1. Place the meat in a roasting pan with the wine, olive oil, the fennel seeds, crushed with a pestle and mortar, and the orange zest, and bake in the oven for 20 minutes at 400°F/200°C.
2. Remove the meat from the oven, pour off the juices and mix them with the honey and paprika. Pour the sauce over the meat and return to the oven for a further 20 minutes.
3. Serve with a side salad of red onions and sliced oranges, dressed with extra virgin olive oil.

Wines: This dish cleverly combines the slight sweetness of pork with the spicy, aromatic flavor of the other ingredients. To go with it, choose a strong, well-structured red wine, with a well-developed bouquet of herbs and red fruits, crisp and delicate with not too much tannin, such as Terre de Franciacorta Rosso, Solaia, or Rapitalà Rosso.

ROBERTO ANDREONI

BROAD BEAN SOUP
WITH SALAMI BALLS AND ARTICHOKES

Serves 4-6

1¼ cups/250 g of dried broad beans, soaked in cold water for at least 6 hours, 1 carrot, chopped, 1 medium onion (chopped), 4 oz/50 g of celery (chopped),
1 potato, diced, 6 artichokes, 1 shallot, 3 bay leaves, 100 g of extra virgin olive oil, 1 lb 5 oz/600 g of salami sausage meat,
3 potatoes, salt and pepper to taste

Method: 1. Brown the carrot, onion, celery and potato in ⅓ of the oil, then add the broad beans with enough water to cover. Boil for about 1 hour.
2. Trim the artichokes, removing the sharp points of the leaves and the hairy choke, and divide into segments. Finely chop the shallot and brown in ⅓ of the oil, add the artichokes and cook, covered, until tender.
3. Meanwhile, shape the sausage meat into balls. Cut the potatoes into matchsticks. Coat the salami balls with the potatoes and bake in the oven with the remaining oil until golden.
4. Purée the broad beans in a mixer, transfer the soup to a tureen, arrange the salami balls and artichokes on top, and serve.

Wines: The succulent, aromatic and slightly sweet favors of this dish are best complemented by a strong, full-bodied, young red wine with a fruity, floral bouquet, delicate and pleasantly acidic, with just a touch of tannin. Good choices include Isonzo del Friuli Franconia, Rosese di Dolceacqua, or Colli Pesaresi Sangiovese.

DARKO BAN, ELVIO MUHA, PAOLO POLLA

TRIESTE-STYLE BOILED PORK

Serves 4

4 pork ribs, 9 oz/250 g of smoked pancetta or streaky bacon,
2 salsicce di cragno (typical smoked, matured sausages from Trieste) or, if not available, fresh pork sausages,
1 cotechino weighing 7 oz/200 g, 1¾ lb/750 g of fresh pork (preferably shoulder or neck), 2 frankfurters, horseradish, mustard
For the sauerkraut:
a generous 2 lb/1 kg of pickled sauerkraut, 4 oz/100 g of pancetta or streaky bacon, 2 oz/50 g of pork fat, caraway seeds, salt and pepper

Method: 1. Bring plenty of water to a boil in a large pot. Add the ribs, the pancetta and the sausages. (If you are using fresh sausages, do not add them yet.)

2. When the water boils, add the *cotechino* and pork, turn down the heat and simmer for 20 minutes. Then add the fresh sausages, if using, and continue to cook for a further 20 minutes. Finally, add the frankfurters and cook for another 5 minutes.

3. To prepare the sauerkraut, finely dice the pancetta and slowly brown it in lard, or, if you prefer, fat from boiled ham. Add the caraway seeds and sauerkraut, season with salt and pepper and moisten with a few ladlefuls of stock from the meat. Cook gently until all the liquid has been absorbed.

4. Arrange the boiled meats on a hot serving dish and, before serving, remember to add a little more salt to the slices of fresh pork. Serve with plenty of grated horseradish, mustard, and a few spoonfuls of sauerkraut.

Wines: The rich, succulent, sweetish flavors of the main ingredients of this typical winter dish contrast with the subtle effect of the horseradish and mustard. As an accompaniment, serve a very delicate, medium-mature, red wine with a rich, fruity, floral and slightly spicy bouquet, with just the right touch of tannin and full-bodied. Darmagi, Solaia, or Regeleaili Rosso Riserva del Conte would be excellent choices.

ENDIVE AND PROVOLA PANCAKES WITH SAUSAGE AND BASIL SAUCE

Serves 4

For the pancake batter: 2 eggs, 1 cup/150 g of flour, 2 cups of milk, lard for greasing the frying pan, salt and pepper
For the filling: 1 head of Belgian endive, 4 oz/100 g of mild provola (buffalo cheese), 4 tsp/20 ml of olive oil, 2 cloves of garlic, 1 tbsp of tomato paste,
shallots, cut into long threads, salt and pepper
For the sauce: 4 oz/100 g of sausage, 7 tbsp of brown stock, 4 basil leaves, 1 clove of garlic, 4 tsp/20 g of butter, 7 tbsp of vegetable broth, 1 tbsp/20 g of flour

Method: 1. Make the pancakes in a suitable frying pan.
2. Prepare the filling, mixing the chicory with the olive oil, garlic, tomato paste, salt and pepper. Cook, covered, in the oven for about 40 minutes at 350°F/180°C.
3. When the chicory mixture is ready, spread it over the pancakes, lay the thinly sliced cheese on top, then bind the pancakes with the shallot threads.
4. For the sauce, make a roux in a frying pan with the butter and flour, add the brown stock and vegetable broth, and the garlic and sausage chopped into pieces. Season with pepper and cook for about 10 minutes. Purée the sauce in a blender, then pass it through a strainer.
5. Cover each plate with a coating of sauce, place the pancake in the center, garnished with a basil leaf.

Wines: Chicory and basil give this dish a touch of freshness, combined with the rich, savory and aromatic flavors of the other ingredients. Together, they call for a well-structured, smooth red wine, pleasantly acidic, with only a touch of tannin, and a fruit and flower bouquet, such as Valle d'Aosta Enfer d'Arvier, Colli Piacentini Pinot Nero, or Torgiano Rosso Riserva.

BOUCHÉES OF PORK IN FILO PASTRY WITH GORGONZOLA FONDUE AND PISTACHIOS

Serves 4

14 oz/400 g of lean pork, 5¼ oz/150 g of pork fat, the whites of 3 eggs, rosemary, 1 clove of garlic, salt and pepper, lemon zest, 1 pack of filo pastry (shredded filo pastry often used in Greek cuisine), ½ cup/100 g butter, 7 oz/200 g gorgonzola, 1 cup/200 g of cream, a pinch of paprika, 1 bay leaf, pistachios

Method: 1. Cut the pork into pieces, mix it with the fat and add the rosemary, garlic, salt and pepper. Purée the mixture in a blender, then pass it through a vegetable mill.

2. Shape the mixture into *bouchées* (little sausage shapes) and fry them for a few minutes. Melt the butter, and dip the *bouchées* in the butter before wrapping them in filo pastry.

3. Meanwhile, prepare the gorgonzola fondue. Mix the cold cheese, paprika, cream, bay leaf, salt, and pepper, then melt very gently over a low heat and continue to cook until reduced by half.

4. Cook the *bouchées* in a frier at 350°F/180°C, or over a medium heat, for a few minutes.

5. Pour the fondue into the center of each plate, set the fried *bouchée* on top and garnish with pistachios and a sage leaf.

Wines: To complement this rich and tasty combination of succulence and sweetness, choose a smooth, well-structured, fresh, crisp, slightly tannic red wine, with an intense, floral and fruity bouquet. Alto Adige Dunkel Lagrein, Chianti Rufina, or Brindisi Negro Amaro would be ideal.

SERGIO BARTOLUCCI

PORK RISOTTO

Serves 4

2 cups/400 g of risotto rice, a generous 2 lb/1 kg of pork bone, 2 salamini (small salamis), 2 bay leaves, 1½ onions, 1 carrot, 1 rib of celery, salt and pepper, a knob of butter

Method: 1. Make a good, rich stock by boiling the bone and salamini, half an onion, the carrot, celery, and bay leaves.
2. Finely chop the whole onion and brown it in a saucepan with the butter. Add the rice and fry for a few minutes until it takes color. Add a little stock, bring to a boil, then continue to add the stock, a little at a time, until the rice is tender. Check the seasoning and serve very hot.

Note: This delicious version of risotto is always an essential part of the feast celebrating the slaughter of the pigs.

Wines: To complement the subtle sweetness of the rice, the pungent flavors of the salamini and the overall richness of the risotto, serve a strong, young, dry, red wine, with a flower and fruit bouquet, crisp, full-bodied and slightly tannic, such as Barbera d'Asti, Elba Rosso, or Pollino Superiore.

SERGIO BARTOLUCCI

PORK SHINBONE

Serves 4

1 whole pork shinbone weighing about 3½ lb/1.5 kg, ¼ cup/60 g of butter, 1 glass of white wine, 2-3 cloves of garlic, rosemary, thyme, salt and pepper, meat broth as required

Method: 1. Spike the whole shinbone with slivers of garlic. Season the meat with salt and pepper and brown it in a frying pan.

2. Moisten with the white wine, add the herbs, and cook slowly in the oven at 350°F/180°C for about 2 hours. Baste frequently with the pan juices, adding more hot stock, if necessary. After about 1 hour, cover the joint with aluminium foil to prevent burning.

3. Slice and serve very hot.

Wines: The delicious sweetness and richness of the meat, combined with the subtle flavor of herbs, calls for a strong but light, young, dry red wine, with a flower and fruit bouquet, delicate, crisp and slightly tannic. Choose a Costozza Cabernet Sauvignon, Montepulciano d'Abruzzo Colline Terramane or San Severo Rosso.

KARL BAUMGARTNER

STUFFED BREAST OF SALT PORK ON A BED OF LENTILS, WITH MIXED VEGETABLES AND NEW POTATOES

Serves 6

1 boned breast of pork, weighing about 1½ lb/700 g, 3½ oz/100 g of salt, 2 oz/3½ tbsp of cane sugar, 2 tbsp saltpeter, 1 tsp peppercorns, coriander, cumin, fennel, mustard seeds, lemon zest, 2 cups/½ l of milk
For the garnish: 10½ oz/300 g of Castellucio or Puy lentils, 1 small piece of pancetta or streaky bacon, 1 bay leaf and 1 sprig of rosemary, 200 g/7 oz of large onions, 1½ lb/600 g of mixed vegetables (baby onions, Swiss chard), Kipfler or Desiré potatoes, boiled, olive oil, salt and pepper to taste, 1 tsp of sugar, ½ cup of vinegar.
For the stuffing: 7 oz/200 g of bread, thinly sliced, 1 red onion, finely chopped, 1½ oz/40 g of spinach, blanched and finely chopped, 4 eggs, herbs (parsley, marjoram, thyme), nutmeg, milk, ¼ cup/50 g of butter, pepper and salt to taste

Method: 1. Using a knife, separate the rind from the meat to create a pocket for the stuffing. Mix the salt, sugar, saltpeter and spices, rub the mixture all over the meat, and leave to marinate for 24 hours.
2. Rinse the meat with water and leave to soak in the milk for 4 hours. Remove the meat from the milk, wash and dry thoroughly.
3. To make the stuffing, soak the sliced bread in a little hot milk, chop and brown the onions, and add the spinach, herbs and nutmeg. Whisk the eggs in a frying pan over a low heat. Squeeze the bread dry, and when the eggs are half cooked add the bread. Stir in the vegetables, season with salt, pepper and nutmeg, blend thoroughly and leave to cool.
4. Fill the meat loosely with the stuffing and put the joint in a roasting bag with the aromatic herbs, seal firmly, place in a pan of hot water and bake at 250°F/110°C for 3 hours.
5. Boil the lentils, pancetta, bay leaf and rosemary in water for ¾ hour.

6. Boil the large onions in a little salted water, then purée them in a blender with a little cooking water. Stew the baby onions with the sugar and vinegar and fry a few onion stalks for a garnish. Shred the Swiss chard and slice the potatoes and sauté both with butter and seasoning.
7. Finally, reheat the lentils with the puréed onions and a little olive oil, season with pepper and a few drops of wine vinegar
8. Remove the meat from the roasting bag and and carve into slices. Serve hot on a bed of lentils. Garnish with the vegetables and potatoes.

Wines: To complement the aromatic flavors of this sweetish, succulent and slightly fatty dish, serve a crisp young red wine with a rich bouquet of red fruits and flowers, smooth, full-bodied, and well-structured with light tannin, such as Colli Berici Tocai Rosso di Barbarano, Elba Rosso, or Savuto.

FRIED PIG'S LIVER GNOCCHI WITH BEETROOT SALAD AND HORSERADISH

Serves 6

5 stale bread rolls, ½ cup/125 ml of milk, 4 eggs, 9 oz/250 g of pig's liver, 1 pig's spleen, 1 oz/30 g of veal suet, 4 shallots, 1 handful of parsley, 1 tsp of marjoram, 2 cloves of garlic, ¼ oz/5 g of chives, lemon zest, salt, pepper and nutmeg to taste, ½-1 cup/50-100 g of breadcrumbs, ½ cup/100 g of butter, 2 tbsp of flour, 1 egg and ½ cup/50 g of breadcrumbs for coating, 2 cups of oil for frying, 4-5 small red turnips or 2 large turnips and a small piece of fresh horseradish. For the sauce: 1 oz/30 g of strong Dijon mustard, 7 tbsp of extra virgin olive oil, 2 tbsp of walnut oil, 3 tbsp of balsamic vinegar, 1 tbsp of raspberry vinegar, a pinch of wild aniseed, salt and pepper, 1 tbsp soy sauce

Method: 1. Slice the bread rolls thinly, soak them in lukewarm milk then mix with the eggs.

2. Put the liver, spleen, suet and seasonings twice through the mincer.

3. Brown the shallots in the butter, blend them with the meat, season lightly with salt, pepper and a little grated nutmeg. Blend with the bread and egg mixture, and add the breadcrumbs, mixing thoroughly,

4. Peel and boil the turnips, cut them into julienne strips. Mix all the ingredients for the sauce and pour it over the turnips.

5. Shape the meat mixture into walnut-sized gnocchi and steam them for 12 minutes. Then coat them, first in flour, then egg, and finally breadcrumbs and fry them in a frier at 350°F/180°C or over a medium heat, for a few minutes.

6. Arrange a little beetroot salad in shallow bowls, topped with 2 gnocchi per person, sprinkled with a little grated horseradish.

Wines: The sweet, juicy texture of the main ingredients of this dish are delightfully balanced by the spiciness of the horseradish. The best accompaniment is a young, full-bodied, very smooth red wine, crisp, but with very little tannin, and a bouquet of herbs, flowers and fruit. Trentino Pinot nero, Lison-Pramaggiore Refosco dal Peduncolo rosso, or Aprila Merlot would be highly suitable.

WALTER BIANCONI

PAN-FRIED PORK RIB CHOPS WITH BEANS IN RED WINE

Serves 4

16 meaty pork rib chops, 2 oz/50 g of lightly smoked pancetta or streaky bacon, 1 clove of garlic, 2 bay leaves, 1 cup/200 g of dried Lamon or borlotti beans, soaked overnight in cold water, 1 rib of celery, sage and rosemary, 1 qt/1 l of red Cabernet, 1 pinch of cinnamon, 7 tbsp/100 ml of brown veal stock, 4 tsp of extra virgin olive oil

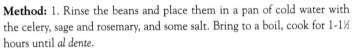

Method: 1. Rinse the beans and place them in a pan of cold water with the celery, sage and rosemary, and some salt. Bring to a boil, cook for 1-1½ hours until *al dente*.

2. In a frying pan, lightly brown the garlic in a little olive oil, add the finely chopped pancetta and the bay leaf; brown the rib chops in the pan, season with salt and pepper and add the beans with a little of the bean stock. Cook for about 20 minutes.

3. Stir in the wine and cinnamon and simmer until most of the liquid has evaporated.

4. Finally, add the veal stock and cook for a further 5 minutes

Wines: Sweet, juicy and spicy with a touch of sharpness, this tasty dish is best served with a medium-mature red wine. Something delicate, rich and full-bodied with a fruity, vegetal, spicy bouquet and not too much tannin, like Cabernet Sauvignon Darmagi, Cabernet or Merlot della Stoppa or Fiorano Rosso, would be most suitable.

WALTER BIANCONI

BOILED PIG'S TROTTER WITH TURNIPS MARINATED IN GRAPE MUST

Serves 4

4 pig's trotters, cleaned and blanched, 1 rib of celery, 1 onion, 2 bay leaves, 4 small red turnips,
7 tbsp of concentrated grape must, 1 horseradish, salt and pepper

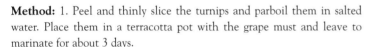

Method: 1. Peel and thinly slice the turnips and parboil them in salted water. Place them in a terracotta pot with the grape must and leave to marinate for about 3 days.
2. Split the trotters lengthwise and bind them with kitchen twine. Place them in a pot of cold water, with the celery, onion and bay leaves. Bring to a boil, skim the scum from the surface, season with salt, and simmer on a low heat for about 4 hours.
3. Bone the trotters and serve them warm with the turnips, topped with grated horseradish.

Wines: The rich juiciness of this dish, combined with a certain sweetness and spiciness, requires a crisp, young, red wine with a well-developed, fruity, floral bouquet, strong but delicate, with not too much tannin. Choose an Alto Adige Santa Maddalena, Pomino Rosso or Rosso di Tursi.

LUCA BOLFO & MARIO ORIANI

MARINATED SLIVERS OF PORK LOIN WITH SWISS CHARD AND BLACK OLIVES

Serves 6

1 lb/500 g of pork, cut from the middle of the loin, 5 cups/750 g of fine salt, 3 cups/750 g of superfine sugar, 1½ qt/1.5 l of water, 20 black peppercorns, crushed, 1 bunch of herbs, consisting of thyme, rosemary, sage, bay leaves, fennel, juniper berries, 3 cloves of garlic, ¾ cup/200 ml of extra virgin olive oil, 4 oz/100 g of pitted olives in olive oil, 3 large heads of Chioggia (Swiss chard), 2 cups/500 ml of white wine

Method: 1. Simmer the water, salt, sugar and some of each of the herbs and seasonings for 5 minutes then leave to cool.
2. Immerse the pork in the marinade and refrigerate it, taking care to turn the meat each day. After 3 or 4 days, drain the meat thoroughly and sprinkle it with the rest of the herbs and seasonings, coarsely chopped, then wrap it in a cloth, previously soaked in plenty of white wine. Refrigerate for 1 day.
3. Finely chop the Swiss chard greens and fry them in a very hot frying pan with a film of oil. Arrange the chard on a serving dish or on individual plates.
4. Slice the meat very thinly, wiping away the herbs, arrange decoratively on the bed of greens and cook for 2 or 3 minutes under a very hot broiler or in a very hot oven. Dress with extra virgin olive oil, freshly ground black pepper and the black olives, drained of their oil.

Wines: This opulent and delicious, slightly oily dish, with its sweetness and flavor of herbs, calls for a strong, young, crisp but delicate and well-structured red wine with just a little tannin. Try Rossese di Dolceacqua, Bosco Eliceo Merlot or Cannonau di Oliena.

| LUCA BOLFO & MARIO ORIANI |

PORK FILLET IN A BUTTERY POLENTA CRUST WITH CABBAGE

Serves 4

A generous pound/500 g of pork fillet in one piece, 10 large cabbage leaves, boiled, 1 qt/1 l of fresh polenta, cooked until tender then kneaded with
½ cup/50 g of butter and 2 oz/50 g of Parmesan cheese, the yolks of 3 eggs, chopped marjoram, 150 g/5 oz of fresh pig's caul fat, washed, 1 glass of white wine,
1 cup/250 g of Mostarda di Cremona (pickled candied fruit), extra virgin olive oil

Method: 1. Lightly coat the pork fillet with flour and fry in a little olive oil for about 20 minutes.

2. Remove the meat and deglaze the pan with the white wine; reduce the pan juices to make a syrupy sauce, then set aside and keep hot.

3. Arrange the ingredients in rectangular layers, in the following order: caul fat, cabbage leaves, polenta, until you have a crust about an inch thick. Place the pork in the center then roll it up so that the meat is completely wrapped in polenta crust.

4. Bake in the oven at 325°F/160°C for about 20 minutes.

5. Cut into four equal slices, arrange them on a serving dish and pour over the sauce. Serve the candied fruits separately.

Wines: The perfect complement to the intense flavors of this tasty dish would be a strong, young, dry, red wine, still quite acidic, but smooth with not too much tannin and plenty of body. Choose Oltrepò Pavese Bonarda, Castel Chiuro Rosso or Costa d'Amalfi Ravello Rosso.

GIANNI BOLZONI

CREMASCA SAUSAGE WITH MUSHROOMS AND POLENTA

Serves 4

1¼ lb/600 g of cremasca sausage, a generous 2 lb/1 kg of field mushrooms, 1 small onion, 1 clove of garlic, 1¾ lb/800 g of ripe, juicy tomatoes, ½ tsp of sugar, salt and pepper, ½ glass of white wine, 1 ladleful of stock, 1 qt/1 l of fresh polenta, cooked

Method: 1. Wipe the mushrooms and rinse them several times in plenty of water with lemon juice.
2. Heat a pan of water and, when it comes to a boil, add the mushrooms, removing them from the pan when they become firm.
3. Meanwhile, put the sausage in a frying pan with half a glass of white wine to boil off the fat.
4. Chop the onion and garlic and brown them in oil. Pass the tomatoes through a sieve and add them to the pan, with a little sugar, white wine and a ladleful of stock. Cook for about 10 minutes. Then add the mushrooms and, after about 20 minutes, the sausage, cut into pieces.
5. Season with salt and pepper and simmer on a low heat for a further 10 minutes. Serve with polenta.

Wines: This quite delicious, rich combination of sweet and spicy flavors comes from the lower Po Valley. To go with it, choose a strong, medium-bodied, young, red wine, maybe even a lightly sparkling one, dry with a fruity bouquet, crisp and delicate, with not too much tannin, such as Dolcetto di Ovada, Colli Piacentini Bonarda or Lambrusco de Sorbara.

GIANNI BOLZONI

PORK RIND AND BORLOTTI BEAN SOUP

Serves 4

7 oz/200 g of pork rind, with as much fat as possible removed, 1 cup/200 g of dried borlotti beans (soaked overnight), 1 onion,
1 lb 5 oz/600 g of ripe tomatoes, ½ cup of oil, ½ glass of white wine, vegetables to make stock (onions, leeks, carrots, celery, Swiss chard,
garlic, bay leaf), basil, salt, pepper, aromatic spices

Method: 1. Make the vegetable stock, then boil the rind in the stock until tender. Remove the rind from the pan and, while still hot, cut into small squares.
2. Boil the beans in water until tender.
3. Meanwhile, prepare a tomato sauce. Chop the onion and peel and chop the tomatoes. Lightly brown the onion in oil, add the white wine, followed by the chopped tomatoes and the basil. Simmer briefly. Add the rind, cook for a few moments, then add the stock. Season with spices.
4. Drain the beans and set aside about a quarter of them, adding the rest to the soup.
5. Purée the reserved beans in a blender and add just before serving. Season with salt and pepper and serve very hot.

Wines: This delicious winter soup is rich and aromatic. It should be accompanied by a smooth, nicely balanced, young red wine, crisp and dry with a fragrant herby bouquet and not too much tannin, such as Monferrato Freisa, Colli Euganei Cabernet or Eloro Rosso.

MASSARIOT'S PICNIC

Serves 6

9 oz/250 g of fresh ricotta, 1 lb/450 g of salami, finely diced, 2 tbsp of butter, 4 oz/100 g of baby salad leaves, 4 oz/100 g of rocket,
extra virgin olive oil and vinegar to taste

Method: 1. Melt the butter in a frying pan, add the diced salami and brown over a high heat for 2 minutes, then add the ricotta and sauté briefly. 2. Arrange a mixture of salad leaves and rocket on each plate, dress with a little oil and 2 or 3 drops of vinegar, and top with 2 spoonfuls of hot salami and ricotta mixture, straight from the pan.

Note: This is a more refined, modern version of an old local recipe. Massariot was a small landowner who was always ready for "a little something".

Wines: To complement the delightful freshness of this appetizer, with its aromatic, slightly sharp flavors, choose a strong and full-bodied, crisp but delicate white wine with an intense bouquet of fruit and flowers, such as Terlano Pinot bianco, Albana di Romagna or Orvieto Classico.

AURELIO BONARDI

ROAST CURED RUMP OF PORK WITH APPLE PURÉE, APPLE FRITTERS AND PRUNES

Serves 6-8

1 piece of cured rump of pork weighing about 2½ lb/1.2 kg, 11 oz/300 g of pancetta or streaky bacon, in rashers, sage and rosemary,
6 Reinette or other crisp eating apples, 1¼ cups/300 g of pitted prunes, 3½ tbsp/50 g of superfine sugar, 2 cinnamon sticks, ½ glass of brandy,
½ glass of white wine, juice of ½ a lemon
For the batter: ¾ cup of beer, 3¼ cups/400 g of flour, 3 eggs, salt and water as required, olive oil for frying

Method: 1. Lard the pork with pancetta, secure with kitchen twine, season with salt and pepper and roast in the oven at 350°F/180°C for 75 minutes.
2. Peel, core and slice four apples, place them in a saucepan with the sugar, brandy, white wine and cinnamon. Simmer over a low heat for about 15 minutes, adding the lemon juice about halfway through. Pass through a sieve, return to the saucepan and simmer until the mixture turns to a purée.
3. Make a batter with the ingredients indicated. Peel, core and slice the remaining apples, then dip the apple slices and prunes in the batter and fry them in hot olive oil for about 3 minutes, until golden.
4. Serve the roast, thinly sliced, with some of the pan juices, garnishing each plate with a few fritters and a spoonful of hot apple purée.

Wines: The delectable aromatic, spicy sweet-and-sour flavors of this rich roast require a well-structured red wine, strong, but not too young, crisp, with a discreet touch of tannin, and an intense bouquet of fruit, flowers and spices. Choose an Alto Adige Blauburgunder riserva, Solaia or Torre Ercolana.

ANNA & LUCIA BOTTE

LAGANE WITH PORK RAGÙ

Serves 4

For the ragù: 1 lb 5 oz/600 g of pork shoulder in one piece, 2 cloves of garlic, 4 oz/100 g of tomato purée, a generous 2 cups/500 g of passata (sieved tomatoes),
1 bay leaf, 2 tbsp of extra virgin olive oil, 7 tbsp of red wine, salt to taste
For the lagane (wide ribbon pasta): 2 cups/ 250g of flour, 7 tbsp of water, 1 tbsp of extra virgin olive oil, 1 pinch of salt
To serve: ½ cup/60 g of grated pecorino piccante cheese

Method: 1. Mix the flour with the water, salt and oil, and knead vigorously for 15 minutes. Leave the dough to rest for 1 hour.
2. Roll the dough into thin sheets and leave to dry for ½ hour, then cut into ribbons 1 in/2.5 cm wide and 7 in/17.5 cm long.
3. Pour the oil into the frying pan, add the garlic, fry until it takes color, then remove it, fry until well browned, then sprinkle with wine and continue to cook until the wine evaporates. Add the tomato purée and passata, the bay leaf and salt, and stir for a few minutes. Add enough water to cover the meat and simmer over a low heat for about 2 hours.
4. Cook the lagane in plenty of boiling, salted water until *al dente*. Serve with the ragù, sprinkled with pecorino.

Wines: The sweetness of the pork contrasts with the pleasantly sharp taste of the tomatoes to create a well balanced blend of flavors. The best choice of wine is a strong, young red, dry, with a fruity bouquet, pleasantly crisp and tannic, such as Rubino di Cantavenna, Friuli-Latisana Franconia or Lizzano Negro Amaro Rosso.

ANNA & LUCIA BOTTE

ESCAROLE SOUP
WITH PORK RIBS OR VERZINI SAUSAGES

Serves 4

16 pork ribs or 8 verzini (pork sausages), 4 heads of escarole (batavia lettuce), 2 tbsp of extra virgin olive oil, 2 cloves of garlic, 3 San Marzano tomatoes, peeled,
¾ cup/100 g grated pecorino, black pepper and salt to taste, 2 ladlefuls of stock

Method: 1. Pour the oil into a frying pan and lightly brown the cloves of garlic.
2. Wash the escarole, cut it into pieces and add it to the pan to wilt the leaves.
3. Chop the tomatoes and add them to the pan with salt, pepper and two ladlefuls of light stock. Boil for about 10 minutes, then add the ribs or sausages and the pecorino. Cook for a further 10 minutes.
4. Serve very hot with piquantly seasoned croûtons.

Wines: This savory starter combines the sweetness of pork ribs with the acidity of tomatoes and the aromatic taste of garlic, pepper and pecorino. To complement it, choose a strong, young, not too highly structured, red wine, with an intense bouquet of fruit and flowers, crisp and delicate, with not too much tannin. Colli Morenici Mantovani del Garda rosso, Val di Cornia Rosso or Alezio Rosso would be ideal.

PORK WITH CAVOLO NERO

Serves 4

2 pork fillets of a total weight of ¾ lb/350 g, 11 oz/300 g of cavolo nero (black Italian cabbage), 1 bunch of wild fennel or fennel seeds, ½ an onion, 2 oz/50 g of sausage, 1 cup of brown stock, 2 cups of meat stock, 2½ tbsp/40 ml of extra virgin olive oil, salt and pepper

Method: 1. Trim and wash the cabbage, and cut it into strips.
2. Chop the onion, and brown it in a frying pan with the sausage. Add the fennel, and finally the cabbage, then stir until the cabbage softens. Add the wine and meat stock and continue to cook over a medium heat for about 20 minutes.
3. Meanwhile, dust the pork fillets with flour. In a separate frying pan, brown the meat in oil. Tip away the oil and moisten the pork with white wine, add the cabbage, brown stock and continue to cook for about 10 minutes.
4. Carve each fillet into four slices and serve on a hot serving dish.

Wines: This typical winter dish offers both a rich blend of delicious flavors and also the interestingly contrasting textures of the pork fillet and cabbage. It calls for a strong, full-bodied medium-mature red wine, with well-developed acidity, low tannin, and a rich bouquet of fruit and flowers. Try Dolcetto di Dogliani, Vin Ruspo de Carmignano or Semidano Mogoro.

SAURO BRUNICARDI

PIG'S LIVER ON BAY TWIG SKEWERS

Serves 4

1 lb 5 oz-1 lb 9 oz/600-700 of pig's liver, pig's caul fat as required, 1 tbsp wild fennel seeds, 4 bay twigs with leaves, 1 red onion, 3 ripe tomatoes,
1 glass of dry white wine, 4 tsp of extra virgin olive oil, salt and pepper

Method: 1. Finely crush the fennel seeds and mix them on a plate with the salt and pepper.
2. Cut the liver into fairly large cubes, roll them in the seasonings and wrap them in the caul fat. Thread the liver onto the bay twigs, alternating the cubes of liver and bay leaves (beginning and ending with a bay leaf).
3. Thinly slice the onion and fry it with a little oil in a frying pan or earthenware pot, then arrange the skewers on top in a single layer. Brown very gently, basting with the wine, and cook until the liver absorbs the wine.
4. Chop the tomatoes and remove the seeds, before adding them to the pan, then leave to cook for about 15 minutes.
5. Serve very hot with cornmeal polenta and white beans with sage. In winter, you can serve it with baby turnips, sautéed with garlic.

Wines: The sweet, aromatic flavors of the liver and herbs are perfectly complemented by a strong, full-bodied but very delicate red wine, crisp but with not too much tannin, with an intense fruit and flower bouquet and plenty of body. Good choices include Colli Tortonesi Barbera, Montecarlo Rosso or Monica de Cagliari Secco.

CLEAR SOUP WITH TAGLIOLINI AND PORK SAUSAGE

Serves 4

7 oz/200 g of salamella (lean, horseshoe-shaped pork sausage), 4 tomatoes, peeled and coarsely chopped, 1 small onion, chopped, 3¼ cups/800 ml of meat stock, 7 oz/200 g of very thin egg tagliolini pasta, a knob of butter, grated Parmesan cheese, a sprinkling of white wine

Method: 1. In a high-sided pan, brown the onion in the butter. Finely chop the salamella and add it to the pan, moisten with a little white wine, then add the tomatoes. Continue to cook until most of the liquid is absorbed.

2. Meanwhile, bring the stock to a boil, add it to the salamella. Add the pasta and return to a boil. Cook the pasta for about 2 minutes.

3. Serve with plenty of grated Parmesan cheese.

Wines: This delicately flavored but delicious and aromatic soup is best complemented by a lightly structured red wine, dry or slightly sweet, perhaps even slightly sparkling, crisp but very delicate, with a bouquet of fruit and flowers and only a touch of tannin. Try Colline Novaresi Bonarda, Lambrusco di Sorbara or Alghero Rosso.

PIG'S TROTTERS AND TAIL WITH CABBAGE

Serves 4

2 pig's trotters, 1 pig's tail, 1 onion, 1 carrot, 1 rib of celery, 2 cloves of garlic, extra virgin olive oil, 1 glass of white wine,
½ a cabbage, salt and pepper

Method: 1. Singe the trotters and tail over a flame to remove any bristles. Wipe thoroughly and place in a pot of boiling salted water with the coarsely chopped onion, carrot, celery and one clove of garlic. Simmer for about 2 hours.
2. Finely chop the cabbage and brown it with the other clove of garlic and olive oil, sprinkle with white wine and cook for about 10 minutes.
3. When the trotters and tail are cooked, add them to the cabbage. Add a little of the meat stock and simmer for a further 10 minutes.
4. Serve with slices of polenta, lightly toasted.

Wines: In this recipe, the sweet, fatty sensation of the trotters harmonizes with the more subtle flavor of the cabbage. Serve a young dry red wine, strong and full-bodied, but delicate, nicely acidic and slightly tannic. Cellatica superiore, Chianti Colli Aretini or Verbicaro Rosso would be good choices.

SERGIO CARBONI

PORK FILLET IN A PASTRY CRUST WITH CARAMELIZED CARROTS

Serves 4

4 pork fillets, each weighing about 4 oz/120 g, 4 slices of prosciutto crudo, ¾ cup/200 ml of brown pork stock, 4 carrots, 1 tbsp of sugar, 2 tbsp of oil, 2 tbsp of butter, salt, pepper, 1 tsp of balsamic vinegar
For the pastry: 2½ cups/300 g of white flour, 1¼ cups/150 g of coarse salt, the white of 1 egg, finely chopped rosemary, sage, thyme, marjoram and parsley, water as required

Method: 1. Quickly brown the pork fillets in a frying pan with a little oil. Remove the fillets from the pan and drain them on kitchen paper. Season with salt and pepper and wrap each fillet in a slice of prosciutto.
2. Knead all the pastry ingredients together to make a smooth dough. Roll out the pastry and cut into rectangles, wrap each fillet in pastry, seal well and brush with egg white. Bake in the oven for 20 minutes at 350°F/180°C.
3. Cut the carrots in sticks, gently rounding the edges (see illustration). Boil them in salted water and then caramelize them in a frying pan with butter, a little sugar, salt and balsamic vinegar.
4. Remove the fillets from the oven. Using a small knife, cut carefully around the crust to make a lid. Remove the meat, slice it, and return it to the pastry case. Arrange the pies on individual plates, pour a

little hot stock over the meat before replacing the lids, and garnish with the carrots.

Wines: This nicely balanced dish combines the sweet and succulent flavor of pork with that of aromatic herbs and the pleasingly tangy garnish. The accompanying wine has to be a mature, well-structured red, strong but delicate, with an intense fruit and flower bouquet, such as Barbaresco, Cialla Schiopettino or Etna Rosso.

SERGIO CARBONI

PORK, CABBAGE AND RICE SOUP

Serves 4

For the broth: pork bone, celery, onion, leek and salt
For the soup: 1 cup/200 g of vialone nano or other risotto rice, 40 oz/100 g of ground fresh pork, 4 oz/100 g of cabbage, shredded,
½ cup/80 g of grated Parmesan cheese, 1 cup/ 80 g of croûtons

Method: 1. Prepare the broth with the ingredients indicated, leaving it to boil for at least 2 hours.
2. Separate enough for four servings (about 1 qt/1 l) and pass it through a fine sieve. Bring to a boil and add the cabbage. Cook for about 5 minutes before adding the rice.
3. When the rice is almost cooked, add the ground pork, mixed with a little of the broth.
4. When the rice and meat are thoroughly cooked, pour the soup into an earthenware tureen, sprinkle with croûtons and grated Parmesan cheese and serve very hot.

Note: This traditional soup from the Po Valley is an essential part of the feast to celebrate the day when the pig is slaughtered.

Wines: This soup combines the sweetness of the rice, pork, and cabbage, combined with the subtle sharpness of the cheese. Serve with a young, dry, full-bodied, red wine of medium strength, fresh and delicate with only a touch of tannin, perhaps Bagnoli rosso, Sant'Antimo Pinot Nero or Sambuca di Sicilia Cabernet Sauvignon.

MARCO CAVALLUCCI

PORK SHINBONE WITH POTATOES AND CELERY

Serves 4

2 pork shinbones, a generous pound/500 g of potatoes, 1 whole head of celery, 1 ladleful of brown stock, 2 cups/500 ml of dry white wine, 2 springs of rosemary, sage, 2 cloves of garlic, ⅓ cup/80 ml of extra virgin olive oil, salt, pepper

Method: 1. With the point of a knife, make an incision along the bones, season with salt, pepper, rosemary, sage and garlic. First brown over a high heat on top of the oven, then bake at 400°F/200°C.
2. Peel the potatoes, cut them into chunks and sauté in very hot oil.
3. Wash and chop the celery and sauté in a little oil.
4. After about 1 hour, pour the wine over the shinbones. When the wine has evaporated, add the potatoes, celery, and brown stock and return to the oven. (Shinbones need between 1½ and 2 hours, according to size.)
5. Carve the meat vertically, and serve on warmed plates with the celery and potatoes.

Wines: The meat and potatoes lend this dish intense sweetness, contrasting with the more subtly aromatic flavors of the other ingredients. Serve with a crisp, not too young, red wine, strong but delicate, with a touch of tannin and plenty of body, such as Valpolicella, Parrina Rosso or Donnici.

OVEN BAKED PORK FILLET WITH CHIVE ZABAGLIONE

Serves 4

8 slices of pork fillet, each weighing about 3 oz/75 g, 1⅔ cups/400 ml of white wine, 1⅔ cups/400 ml of white vermouth, 2 cups/500 ml of meat stock, extra virgin olive oil, salt and ground black pepper
For the zabaglione: the yolks of 8 eggs, 1 shallot, chopped, 1 cup/250 ml of white wine, ¾ cup/200 ml of white vermouth, chives, chopped, salt and ground black pepper

Method: 1. Season the meat with salt and pepper and brown briefly in oil in a very hot frying pan.
2. Bring the brown stock to a boil with the wine and vermouth.
3. Place the meat in a roasting pan, cover with the stock and bake in the oven at 400°F/200°C for 6-8 minutes. Remove the meat from the oven and cut into slices.
4. To make the zabaglione, mix the chopped shallot, vermouth and wine, and reduce over a medium heat. Remove from the heat and stir in the egg yolks. Cook in a double boiler, adding a little meat stock. Beat with a fine whisk to make a frothy zabaglione. Check the seasoning and finally add the chopped chives.
5. Arrange the meat on warm plates, coat with zabaglione, and garnish with bunches of chives.

Wines: The sweet and succulent pork fillet contrasts with the discreet flavors of the other ingredients. This dish calls for a strong, well-structured, mature, white wine, pleasantly acidic with a well-developed and rich bouquet of fruit and flowers, such as Isonzo del Friuli Riesling Renano, Capezzana Bianco or Bianco d'Alcamo.

POTTED PORK FAT TUSCAN-STYLE

Ingredients:

2 lb/1 kg pork fat, 4 tsp/20 g sea salt, ground black pepper to taste, 5-6 cloves of garlic, a few sprigs of rosemary, a few drops of strong wine vinegar

Method: 1. Take good, firm pork fat from the back of the pig and grind it very finely. Mix with the sea salt and pepper, crushed in a pestle and mortar, a few drops of strong vinegar, and finely chopped rosemary.
2. Knead the mixture on a marble surface (this is essential), until soft, light and creamy.
3. Potted pork fat should be eaten, spread on bread still hot from the oven, with a good Chianti Classico or other first-class wine. It will keep for some time.

Note: As you eat this simple spread, also known as Crema del paradiso, remember that it is not a recent invention, designed to please the connoisseur. Generations of peasants and poor town-dwellers have enjoyed it as a nourishing and comforting meal.

Wines: This recipe is as simple as it is delicious, its rich, fatty texture enhanced by the other aromatic ingredients. To drink with it, choose a strong, young, well-structured red wine, crisp but delicate, and lightly tannic, such as Coste della Sesia Vespolina, Chianti Colli dell'Etruria Centrale or Penisola Sorrentina Lettere Rosso.

DARIO CECCHINI

BEAN SOUP WITH PICKLED PIG'S TROTTER

Serves 8-10

1 pig's trotter (pickled in vinegar, salt, pepper, garlic, bay leaves), rosemary,
2 cups/500 g of zolfini or other dried beans, 1-2 handfuls of Swiss chard, cabbage or other cooked greens,
1 cup/200 g of spelt, or 1¼ cups/300 g short pasta of your choice, 8-10 slices of Tuscan or other coarse country bread,
salt, a few cloves of garlic

Method: Pickled pig's trotters are prepared as follows: 1. Take trotters from newly slaughtered pigs. Slit each one open lengthwise with a knife, leaving the top part attached, so that the trotter stays in one piece. Marinate in a solution of equal parts of water and vinegar, and some bay leaves (at least one per trotter), and leave for several days in a cool place.

2. Stack the trotters in layers, seasoned with crushed garlic, crumbled fresh rosemary leaves and sea salt. After a week, rinse them again in water and vinegar, sprinkle generously with pepper and hang to dry. Before using a trotter, soak it in water for about 24 hours.

3. To make a good bean soup (although you can use pig's trotters in an infinite number of ways, this is the most usual), you will need one trotter for just under 2 cups (450 g) of beans (which must first be soaked in water overnight). Use plenty of water to make the soup, originally a very humble dish, meant to feed the largest possible number of people. Cook for about 2 hours. There is no need to add salt or any other seasoning.

4. Strip the meat off the bone and cut it into small pieces. Pass some of the beans through a sieve to thicken the soup. Add some seasonal greens (Swiss chard, cabbage, etc) and, if you like, spelt or pasta. You may also place a slice of toasted country bread, rubbed with garlic, in the bottom of each soup bowl and pour the very hot soup on top.

Wines: This is a gloriously tasty example of peasant cuisine, with a rich blend of sweet, savory and aromatic flavors. Serve with a strong, full-bodied, young red wine, with a fruity bouquet, delicate but nicely acidic with a touch of tannin. Try Montello e Colli, Asolani Cabernet Franc, Colli Altotiberini Rosso or Rosso de Cerignola.

FILIPPO CHIAPPINI DATTILO

SPICY GLAZED SMOKED LOIN OF PORK WITH BRAISED CABBAGE AND MOSTARDA

Serves 6

1 smoked (or cured) loin of pork weighing about 2 lb/1 kg, 4 tbsp of honey, 2 tbsp of Dijon mustard, 1 cup/300 g of mostarda (pickled candied fruits), 2 cups/250 g butter, 1 medium cabbage, 2 tbsp of sugar, 1 tbsp of finely ground spices (fennel seeds, red peppercorns, coriander seeds and juniper berries), 2 tbsp of brandy, 2 glasses of dry white wine, 2 cloves of garlic, 1 sprig of rosemary

Method: 1. Brown the meat in a little butter with garlic and rosemary. When it has browned all over, pour in the white wine, and continue to cook until the wine has evaporated.

2. Mix the honey, spices, mustard and brandy and brush the meat with the mixture, then place it in a preheated oven at 350°F/180°C and roast for about 45 minutes. Baste frequently with the pan juices, brushing the surface of the meat to create a good glaze.

3. Finely chop the cabbage and braise it in the rest of the butter for about 1 hour, with 2 tbsp sugar, salt, pepper and a few juniper berries.

4. Serve the meat very hot on a bed of cabbage. Garnish the serving dish with diced mostarda.

Wines: Sweet, spicy flavors predominate in this rich, succulent and aromatic dish. To harmonize with them, choose a strong, elegantly structured red wine with an intense, well-developed bouquet of herbs, flowers and spices, crisp and very delicate, with just the right amount of tannin. Teroldego Rotaliano, San Giorgio (Cabernet Sauvignon) or Corvo Duca Enrico would all be excellent.

PIG'S TROTTER ROULADE WITH TRUFFLES AND SAUCE PÉRIGUEUX

Serves 4

4 pig's trotters, 14 oz/400 g of turkey, 2 rabbit livers, 7 oz/200 g of chicken livers, 11 oz/300 g of mushrooms, 1½-2 qt/1.5-2 l of vegetable stock,
1 tbsp of chopped parsley and garlic, 2 eggs, 1 tbsp of port, 1 tbsp of brandy, 7 oz/200 g of pig's caul fat, 1½ oz/40 g of black truffles, salt, pepper and nutmeg
For the sauce: ½ cup of brown stock, 2 tbsp of port, 1½ oz/40 g of black truffles

Method: 1. Boil the trotters in plenty of vegetable stock for about 2 hours until the meat falls off the bone, then leave to cool.
2. Meanwhile, prepare the roulade mixture. Finely dice the turkey and the rabbits' and chicken livers and fry them gently. Chop the mushrooms and, in a separate pan, brown them in a little butter with the chopped garlic and parsley. Allow all the ingredients to cool, then mix them together. Season with salt, pepper and nutmeg, adding the port and brandy and binding with the egg.
3. Bone the trotters and dice the meat and mix it with the other ingredients.
4. Take 8 pieces of caul fat and place about 2 tbsp of the mixture on each, top with a few slivers of truffle then wrap the caul around the filling.
5. Fry in a little oil flavored with rosemary until golden all over.

Wines: The combination of the richness of the pork, the sweet and spicy flavors of the other ingredients, and the slightly bitter aftertaste demand a strong, young, delicate red wine, crisp, with not too much tannin, and a fruity, floral bouquet. Try Ghemme, Chianti Classico or Regaleali Rosse del Conte.

6. To make the sauce périgueux, mix the brown stock, port and finely chopped truffles in a small pot over a low heat. Serve the roulades sliced on a bed of potato purée, covered with the sauce.

PIG'S EAR SALAD

Serves 4

6 pig's ears weighing a total of 1 lb 10 oz/800 g, 4 cups/1 kg of goose fat, 4 tsp of butter, parsley, chives, mint, chervil, dill, radicchio, lamb's lettuce, escarole (batavia) lettuce, 2 tbsp balsamic vinegar, 3 tbsp of extra virgin olive oil, 2 cloves of garlic, pepper, thyme, bay leaf, coarse salt

Method: 1. The day before you need them, clean the pig's ears and singe them over a flame to remove any bristle. Put them in a large pot, cover with cold water and bring to a boil. Drain off the water and dry the ears on kitchen paper.
2. Preheat the oven to 475°F/250°C. Cut the ears in half lengthways and arrange them in a roasting pan. Sprinkle with coarse salt, thyme, crumbled bay leaves and a chopped clove of garlic. Place the pan in a double boiler, cover with baking foil and bake in the oven for 45 minutes. The following day, remove all the fat from the ears and cut them lengthways into julienne strips.
3. Wash the salad leaves, finely chop all the herbs, and make a vinaigrette with oil, vinegar, salt and a crushed clove of garlic. Toss and season the salad and divide between four plates.

Wines: This delicious recipe is a wonderful combination of the traditional, succulent taste of pork with the fresh, herby flavors of the salad. Serve with a mature, crisp, dry, white wine, with an intense bouquet of exotic fruits, spices and flowers, fairly alcoholic and full-bodied, such as Alto Adige Terlano Riesling Renano, Colli Martani Grechetto di Todi or Bianco d'Alcamo.

4. In a non-stick frying pan, briefly sauté the ears in the butter over a high heat, then arrange them on a bed of salad, topped with a little freshly ground black pepper.

| GEORGES COGNY |

PORK SHINBONE WITH A CITRUS GLAZE AND RED CABBAGE SALAD

Serves 4

2 fresh pork shinbones, 4 lb/2 kg of coarse salt
For the garnish: 2 carrots, 2 ribs of celery, 1 leek, 1 bouquet garni, 3 tsp of pepper, 6 cloves, 6 juniper berries, 1 tsp of coarse salt
For the sauce: 2 grapefruits, 2 limes, 2 lemons, 4 oranges, 3 tbsp of honey
For the salad: 1 small red cabbage, 2 tbsp of vinegar, 1 tbsp of white wine, extra virgin olive oil, salt and pepper

Method: 1. Marinate the shinbones for 2 days in coarse salt, then wash and dry them.

2. Cut the cabbage into julienne strips, place in a bowl with 1 tbsp of vinegar, cover with cellophane film and leave to stand overnight. This gives the cabbage a good strong color.

3. Place the shinbones in a deep saucepan, with the vegetables and seasonings. Cover with cold water, and cook gently for 10 hours.

4. Preheat the oven to 275°F/140°C. Squeeze the juice from all the citrus fruits and mix them with a ladleful of the pork stock. Place the shinbones in a roasting pan in the oven. Brush the meat with honey and sprinkle with some of the citrus juice, then turn the meat over and repeat the process, reserving half of the juice. Caramelize in the oven for 45 minutes.

5. Transfer the meat to a serving dish and deglaze the pan with the remaining citrus juice. Drain the cabbage and toss in the salad dressing. Serve the shinbones cut in half, sprinkled with citrus sauce and accompanied by the salad.

Wines: The choice of wine to complement this dish is dictated by the intense flavor of citrus fruits and the other aromatic ingredients. The best choice would be a very delicate, young, red wine, fairly acidic with just a touch of tannin and a rich bouquet of fruit, flowers and spices. Try Valle d'Aosta Enfer d'Arvier, Colli di Luni Rosso or Solopaca Rosso.

WARM PORK SALAD WITH CANDIED FRUITS AND MIXED VEGETABLE SALAD

Serves 4

1 pork fillet weighing about 7 oz/200 g, 1 oz/30 of pig's caul fat, 4 oz/100 g of prosciutto crudo, 2 cloves, ground, 1 bay leaf, crumbled, salt and pepper
For the salad: ½ a carrot, 1 rib of celery, 2 leeks, ⅓ cup/80 g of candied citron and orange, 1 heaped tbsp/20 g of sultanas, 3 tbsp/20 g of pine nuts,
4 chestnuts, roasted, 2½ tbsp of Brisighella (or other superior quality) extra virgin olive oil, a few drops of balsamic vinegar

Method: 1. Season the pork fillet with crumbled bay leaf, salt and cloves. Wrap it first in *prosciutto* and then in caul fat. Place the fillet in a copper pan and roast in the oven at 425°F/220°C for 7 minutes, then remove from the oven and keep warm.
2. Lightly steam the vegetables, then arrange them on four plates, with the candied fruits and pine nuts, the coarsely chopped chestnuts, and sultanas.
3. Very gently warm the olive oil. Thinly slice the pork fillet and arrange it on top of the bed of salad, sprinkled with a few drops of balsamic vinegar and the warm olive oil.

Wines: This dish creates a delicious balance between the sweetness of the meat and *prosciutto*, the discreet tang of the balsamic vinegar, the aromatic flavors of the spices and candied fruit, and the richness of the oil. Serve with a mature and well-structured white wine, crisp and smooth with a well-developed bouquet, such as Chardonnay Ronco del Re, Fiorano Bianco or Costa d'Amalfi Ravello Bianco.

POLENTA SAVARIN WITH PORK AND SAUSAGE MOUSSE AND TAGLIONINI

Serves 4

For the savarin: 7 tbsp of polenta, 1¼ cups/300 ml of milk, 1⅔ cups/400 ml of water, 2 tbsp of extra virgin olive oil, salt to taste, 4 individual savarin molds.
For the mousse: 8 pork rib chops, 3 tbsp of celery, chopped, ¼ cup of carrot, chopped, 1 small onion, chopped, 1 clove of garlic, chopped, 4 oz/100 g of pork fat,
7 oz/200 g of pork sausage, 4 button mushrooms, 1 glass of white wine, 2 tbsp of toasted flour, 2 tomatoes, peeled and deseeded, 1 tbsp tomato purée,
chicken stock as required, salt and pepper to taste
For the tagliolini: 1 cup/130 g of flour, sifted, the yolks of 4 eggs, 2 tsp of squid ink.
To garnish: 2 oz/50 g of bone marrow, parboiled, 36 slivers of white truffle, parsley

Method: 1. For the savarin: In a deep saucepan bring the water to a boil with the milk, olive oil and salt. Pour the polenta into the pan in a single stream, and boil on a very low heat for about 40 minutes, stirring constantly. Grease the savarin molds with olive oil. Fill the molds with the cooked polenta and set aside.

2. To make the pork mousse: In a frying pan, soften the celery, carrot, onion and garlic in the pork fat. Skin and chop the sausage, and add to the pan with the rib chops and whole mushrooms. Add the toasted flour and the white wine. When the wine has evaporated, add the chopped tomatoes, tomato purée and some stock. Simmer over a very low heat for about 2 hours, adding more stock, if necessary. Strip the meat from the chops, then pass all the ingredients through a blender.

3. To make the tagliolini: Blend the flour with the egg yolks and squid ink. Knead for about 5 minutes to make a smooth, firm dough. Roll out

the dough into a very thin sheet, then roll it up and cut it into thin strips with a knife. Cook the tagliolini in plenty of lightly salted water until *al dente*.

4. Turn the savarins out onto individual plates and coat them in the pork mousse. Fill the center of each one with tagliolini, dressed with olive oil. Garnish with slices of parboiled bone marrow and parsley and sprinkle with slivers of truffle.

Wines: This appealing combination of flavors, from the mildness of the polenta to the spicy, aromatic tang of the sauce, calls for a strong, full-bodied young red wine, crisp with a rich fruity bouquet and just a touch of tannin. Choose Sandbichler Rosso, Colli Martani Sangiovese or Matino Rosso.

PORK MEATBALLS WITH SAVORY POTATO PURÉE

Serves 6

For the meatballs: 11 oz/300 g of finely ground pork shoulder,
11 oz/300 g of finely ground pork fat, salt, pepper, fennel seeds (optional), 2 tbsp of vinegar, 1 knob of butter
For the purée: 2 lb/1 kg of floury potatoes, 4 oz/100 g of pancetta or streaky bacon, 4 oz/100 g pork fat, salt, pepper, oil

Method: 1. Shape the ground meat and fat into small balls, seasoning lightly with salt and pepper, and adding a few fennel seeds (optional). Brown in a frying pan, then deglaze the pan with a little vinegar. Add a little butter and melt slowly to create a sauce to serve with the meatballs.
2. Boil the potatoes, then peel and purée them.
3. Cut the pancetta or streaky bacon and the pork fat into thin strips, brown in the frying pan then stir into the potato purée. Season with salt and pepper and add a little extra olive oil to create a softer purée.
4. Arrange the purée in an attractive shape on each plate and serve with the meatballs and buttery sauce.

Wines: This simple but rich and savory dish with its strongly flavored ingredients requires a crisp, young, possibly lightly sparkling, red wine, with a rich, fruity bouquet, well-structured, strong and not too soft, with plenty of tannin. Try Freisa d'Asti superiore, Colli di Luni Rosso or Ligorio Rosso.

ENZO DE PRÀ

SWEET BLACK PUDDING

Serves 8

¾-1¼ cups/200-300ml of very fresh pig's blood, not coagulated, 1 lb/500 g of apples, peeled and sliced, 1 cup/100 g of very fine flour, 2 whole eggs, 12 tbsp of milk, 3½ tbsp/50 g of sugar, ¼ cup/50 g of raisins, 3 dried figs, chopped, ½ sachet of baking powder, 1 tsp cinnamon, butter and salt

Method: 1. Whisk the eggs in a bowl and add the milk, blood, cinnamon, a pinch of salt, baking powder, figs, apples and raisins. Stir slowly with a wooden spoon, and finally blend in the flour, continuing to stir until the batter is smooth with no lumps.
2. Butter and flour a 12 in/25 cm cake tin, fill it with the batter and bake in the oven at 300°F/150°C for 30 minutes. The black pudding can be served warm, but it is just as good eaten cold the next day.

Wines: This traditional country cake is rich, sweet and spicy. The best accompaniment would be a young, delicate, lightly structured red wine, fairly acidic with very little tannin, and a rich, fruity bouquet, such as Lago de Caldaro Auslese, Colli Orientali del Friuli Cialla Schioppertino or Forastera d'Ischia.

ENRICO DERFLINGER

CABBAGE PARCELS FILLED
WITH PEARL MILLET AND PORK

Serves 4

4 cabbage leaves, 1 egg, ⅓ cup/80 g of pearl millet, 4 oz/100 g of lean pork, 4 slices of Parma ham, ½ shallot, chopped, ⅓ cup/40 g of grated Parmesan cheese,
2 Belgian endives, 2 oz/50 g of finely diced carrot and celery, ½ tsp of fennel seeds, 2½ tbsp of extra virgin olive oil, ½ cup of vinegar, salt and pepper

Method: 1. Dicè the pork into small pieces, season with salt and pepper and brown in hot oil in a frying pan.
2. Boil the millet for about 20 minutes in water seasoned with fennel seeds. Drain and season with salt, pepper and Parmesan cheese, then stir in the egg.
3. Brown the diced carrot, celery and shallot in a little oil, then add the diced pork.
4. Meanwhile, blanch the cabbage leaves for a few seconds in boiling water and vinegar. Spread the leaves out on a chopping board, lay the slices of Parma ham on top, and then add the pork and millet mixture and fold over to make parcels.
5. Brown the parcels under the broiler, then cut them into diagonal slices. Serve with braised Belgian endives arranged in a star shape on the plate.

Wines: The subtly sweet and aromatic flavors of this northern Italian specialty demand a delicate but very crisp and well-structured, dry, white wine, with a good level of alcohol and an intense bouquet of fruit, flowers and spices. Good choices would be Alto Adige Terlano Riesling Renano, Torgiano Chardonnay or Menfi Feudo dei Fiori.

ROAST LEG OF PORK WITH ROAST POTATOES AND TOMATOES AU GRATIN

Serves 6

1 knuckle end of a leg of pork weighing 3-3½ lb/1.4-1.5 kg, 2-3 cloves of garlic, 1½ tbsp of chopped wild fennel, ½-¾ tsp of black pepper, 2 tbsp of salt
For the roast potatoes: 2 cloves of garlic, ¼ cup of extra virgin olive oil, 1½ lb/600 g of potatoes, black pepper to taste, a sprig of rosemary, 2 tsp of salt
For the tomatoes au gratin: 6 tomatoes, 1-2 cloves of garlic, crushed, 2 tbsp of extra virgin olive oil, ½ cup/50 g of breadcrumbs,
1-2 tbsp of chopped parsley, salt to taste

Method: 1. Clean and bone the leg of pork. Chop the garlic, grind the pepper and mix with the salt and chopped fennel. Rub all the seasonings into the meat, continuing for several minutes, to allow the flavors to penetrate. Leave to macerate in the refrigerator for at least 8 hours.
2. Rub a little more salt into the rind, then make a series of diamond-shaped incisions about 2 in/5 cm along each side.
3. Preheat the oven to 475°F/240°C and roast the meat for 20 minutes, then reduce the temperature to 375°F/190°C and cook for 1¾-2 hours.
4. To prepare the potatoes: Peel the potatoes and cut them into 1 in/ 2.5 cm cubes. Arrange them in a roasting pan and add the seasonings. Roast in the oven with the pork at 375°F/190°C for at least 40 minutes.
5. To prepare the tomatoes: Wash them, cut them in half crosswise and remove the seeds. Wash and finely chop the parsley, peel and finely chop the garlic. Mix the breadcrumbs with 1 tbsp of the olive oil and the chopped parsley and garlic. Arrange the tomato halves in a roasting pan and sprinkle them with the remaining oil, salt and pepper. Spread a little of the breadcrumb mixture on the top of each one and bake in the oven at 375°F/190°C for at least 50 minutes.
6. Serve a generous portion of meat with at least 10 per cent crackling, with roast potatoes and 2 halves of tomato au gratin, garnished with a celery leaf.

Wines: This deliciously succulent and aromatic dish should be served with a mature, full-bodied, dry, red wine, with an intense bouquet of fruits, flowers and spices, very crisp and delicate with a touch of tannin and plenty of alcohol. Boca, Barco Reale di Carmignano or Sagrantino de Montefalco would be ideal choices.

PORK FILLET
WITH ZUCCHINI AND POTATO CAKE

Serves 12

3 lb/1.35 kg of pork fillet, 5 oz/120 g of rolled pancetta or streaky bacon, garlic, bay leaves, sage, rosemary, a pinch of Madras curry powder, 7 tbsp of vinegar
For the zucchini sauce: ⅔ cup of water, 2 tsp of crushed garlic, 2 tsp of freshly chopped wild fennel, 6 tbsp of vegetable oil, ⅓ cup of extra virgin olive oil,
3 oz/90 g of potatoes, 1 tbsp of salt, 6 oz/150 g of zucchini
For the potato cakes: 1¼ lb/600 g of potatoes, 6 oz/150 g of white onion, 5 oz/130 g of pig's cheek, ½ cup of single cream, salt and pepper to taste
For the zucchini: 1 lb/450 g of zucchini, ⅔ cup/150 ml of water, 1 tsp of crushed garlic, 1 tsp of freshly chopped wild fennel, 5 tbsp of extra virgin olive oil, 1 tsp salt

Method: 1. Rinse and chop the herbs. Wash the pork fillet and season it with the herbs and spices, then lard it with the pancetta, making sure that it is firmly attached to the meat. Brown the fillet in a very hot frying pan, and pour over the vinegar.
2. Transfer to the oven and roast at 300°F/150°C for about 2¼ hours, then leave to rest in the cooling oven for at least 5 minutes. Carve into 1 oz/30 g slices.
3. For the zucchini sauce: Boil and peel the potatoes. Slice and steam the courgettes until tender. Transfer all the sauce ingredients to a blender and grind at high speed to create a creamy, frothy sauce.
4. To make the potato cake: Peel the potatoes and onion, slice the potatoes into very thin rounds and the onions into fine julienne strips. Thinly slice the streaky bacon. Line a cake tin with the bacon, arrange the potatoes and onion in layers, season with salt and pepper and add the cream. Bake in the oven at 350°F/180°C for about 1 hour (or longer if the potato cake is baked at the same time as the meat), covering the tin with foil if the surface browns too quickly.
5. Wash and thinly slice the zucchini, season and sauté in olive oil with the garlic and fennel until tender and lightly browned.
6. Serve the meat in slices with the sauce, accompanied by thin wedges of potato cake and the zucchini.

Wines: This delicate, aromatic and extraordinarily tasty dish is best complemented by a young, red wine, soft and pleasantly acidic, with not too much tannin, strong and full-bodied with a fruity, floral, bouquet, such as Canavese Nebbiolo, Tignanello or Ischia Piedirosso.

CORRADO FASOLATO

TERRINE OF LIGHTLY SMOKED PORK WITH GOAT'S CHEESE AND THYME SAUCE

Serves 6

2 lb/1 kg of pork loin, 1 oz/25 g of gelatine, 9 oz/250 g of egg whites, 2 cups/500 g of cream, pork fat for greasing the terrine dish, salt and pepper
For the marinade: 4 lb/2 kg of coarse salt, 4 cups/1 kg of sugar, 1 tsp/4 g tenderizing salt
For the sauce: 11 oz/300 g of goat's cheese, 2 cups of milk, 2 cups of cream, 2 sprigs of fresh thyme
To garnish: ¾ cup of tomatoes, peeled, deseeded and finely diced, thyme-flavored olive oil, 6 oz/150 g mixed lollo bianco and lollo rosso leaves

Method:
1. Mix the marinade ingredients and rub them into the meat. Leave to marinate in the refrigerator for about 12 hours. Remove the meat from the marinade and smoke it for about 2 hours.
2. Chill the mixer bowl in the freezer before grinding the pork with the egg whites and cream, salt and pepper. Soak the gelatine in hot water until it dissolves, then add it to the meat mixture.
3. Line a terrine dish with greaseproof paper and then with thinly sliced pork fat before filling with the meat mixture. Bake in the oven at 350°F/180°C for about 1½-2 hours, allow to cool then chill in the refrigerator for several hours.
4. Shortly before serving the terrine, prepare the sauce. Dissolve the goat's cheese in the milk and cream in a saucepan over a very low heat, season with thyme and leave on the heat to reduce to a thick, creamy consistency. Add salt and pepper at the end.
5. Spread the warm sauce over each plate, set a slice of terrine in the center, garnish with diced tomatoes and a few salad leaves, and sprinkle with thyme-flavored oil.

Wines: This classic appetizer of French origin combines the sweetness and slight fattiness of the main ingredients with the aromatic effect of smoking the pork and the addition of thyme. Serve with a full-bodied, young, dry, white wine from aromatic grapes, crisp and discreetly alcoholic, with a rich fruity, floral and spicy bouquet, such as Alto Adige Traminer Aromatico, Fiorano Semillon or Santa Margherita di Belice Catarratto.

HERBY LEG OF PORK WITH ZUCCHINI VINAIGRETTE AND FOIE GRAS-STUFFED POTATOES

Serves 14

1 leg of pork weighing 14-15½ lb/6-7 kg, 2 cups/500 g of fine salt, 1 tsp of ground white pepper, 4 oz/100 g of rosemary, 2 oz/60 g of chives, 1 oz/30 g of thyme,
1 oz/30 g of dill, 2 cups of extra virgin olive oil, 15 potatoes each weighing 3 oz/90 g, 7 oz/200 g of foie gras, 11 oz/300 g of breadcrumbs,
1 cup/100 g of sesame seeds, 1 cup/100 g of poppy seeds, 4 eggs, 1¼ cups/300 g of butter, 1 cup/100 g of flour, 5 oz/150 g of curly endive or frisée,
5 oz of lollo rosso, a generous pound/500 g of zucchini
For the sauce: a generous pound/500 g each of sweet red and yellow peppers (finely diced), 1 cup of balsamic vinegar, 1 leek, 2 carrots, 1 onion,
3½ tbsp of cornstarch, 2¾ cups/700 g of sugar, 2 tbsp of glucose

Method: 1. Bone the pork and reserve the bone. Chop the aromatic herbs and mix them with salt. Rub the meat with the herbs, wrap loosely in foil and roast for 8-9 hours at 350°F/180°C.

2. Parboil the zucchini, leaving them whole, and set aside. Steam the potatoes, then peel and halve them and scoop the flesh from the middle.

3. Brown the *foie gras* in a frying pan, whisk it in a blender with half the breadcrumbs and two of the eggs, then stir in potato flesh. Fill the potatoes with the *foie gras* mixture and put the two halves back together.

4. Beat the remaining eggs, and mix the remaining breadcrumbs with the sesame and poppy seeds. Dry the potatoes on kitchen paper, dip them in the egg then the breadcrumbs, then deep fry them in plenty of oil.

5. Chop the leek, onion and one carrot and brown in a large saucepan.

6. Roast the pork bone, then add the vegetables and cover with cold water. Bring to a boil, and continue to boil until reduced by half.

7. In another saucepan, caramelize 1¼ cups/300 g of sugar, then add the balsamic vinegar. Strain the stock, then add it to the caramelized sugar. Finely dice the peppers and the remaining carrot and stir them into the sauce. Thicken the sauce with the cornstarch.

8. Use remaining sugar and glucose to make spun sugar for a garnish.

9. Thinly slice the zucchini and arrange them in a circle on the serving dish. Arrange the sliced meat in the center, and top with slices of stuffed potato. Glaze with the sauce and garnish with spun sugar and salad leaves.

Wines: The enormously varied ingredients of this dish work together to give a wonderful blend of savory, sweet and aromatic flavors. To go with it, choose a mature, strong but very delicate white wine, with good acidity and plenty of body, such as Collio Sauvignon, Verdicchio de Matelica or Nasco di Cagliari Secco.

LOIN OF PORK WITH STEWED LEEKS WRAPPED IN CABBAGE LEAVES

Serves 4

A generous pound/500 g of mature tenderloin of pork, 14 oz/400 g of leeks, shredded, 4 cooked cabbage leaves, ½ cup/150 g of butter, ⅓ cup/800 ml of dry white wine, rosemary, sage, salt, pepper, stock as required

Method: 1. Sauté the tenderloin in 3 tbsp of butter with sage, rosemary, salt and pepper, until nicely golden. Moisten with 2 tbsp of wine. Once the wine has evaporated, add the stock. Roast in the oven for 15 minutes at 350°F/180°C, then remove from the oven and leave to rest in a warm place for 20 minutes. Carve the meat into four equal pieces.
2. Meanwhile, prepare the leeks. Melt another 3 tbsp butter in a saucepan, add the leeks and fry until soft. Season with salt and pepper and add 3 tbsp wine and cook until the leeks have absorbed the wine.
3. Spread out the cabbage leaves, remove any tough stalks. In the center of each leaf, make a bed of leeks, lay the meat on top, and spread with more leeks. Wrap the meat tightly in the leeks and arrange on a buttered baking sheet.
4. Bring the meat juices to a boil to reduce as required and bind with the remaining butter. Keep warm in a double boiler.

5. Brown the cabbage parcels in the oven for a few minutes at 400°F/200°C. Cut into slices, arrange on individual plates and cover them with the sauce.

Wines: The leeks and herbs in this dish lend it a delicate flavor, providing a perfect balance with the richness of the meat. Serve with a strong, full-bodied, young, red wine, crisp but delicate with only a touch of tannin, such as Lago di Caldaro classico, Lison Pramaggiore Refosco dal Peduncolo Rosso or Fiorano Rosso.

SUCKLING PIG WITH MUSTARD, PEPPERS AND QUINCE JELLY

Serves 4

2 lb/1 kg of suckling pig (from a pig weighing about 20 lb/9 kg), sliced, 4½ oz/120 g of mustard (with whole seeds),
350 g/12 oz of quince jelly, 6 oz/160 g of sweet red peppers, cut into strips (6 strips per serving),
2 lb/1 kg of lard, ½ glass of white vinegar, ½ glass of dry white wine, ¼ cup/50 g of butter, 1½ cups/350 ml of stock (i e. ⅓ of the quantity of lard),
bouquet garni with bay leaf, sage and rosemary, salt and pepper

Method: 1. Place the slices of meat in a stewpot with the lard and stock, and the bouquet garni (wrapped in muslin), salt and pepper. Simmer on a very low heat for about 1 hour, making sure that the lard does not start to fry (if it does, add more stock). When the meat is cooked, drain off the fat.
2. Brown the pork in a pan with butter and rosemary. Spread with mustard and pour over the vinegar. When the vinegar has evaporated, add the wine and peppers. Cook in the oven at 400°F/200°C, until the crackling is crisp.
3. Serve garnished with peppers and quince jelly.

Wines: In this dish, the aromatic sharpness of the mustard and the quince harmonize perfectly with the sweetness and succulence of the pork. Serve with a strong, full-bodied, fairly mature, red wine, crisp, with a little tannin and a bouquet of ripe fruits, flowers and spices. Ideal choices would be Pinerolese Doux d'Henry, Vino Nobile de Montepulciano and Torre Quarto Rosso.

BLADE SHOULDER OF PORK WITH WHITE WINE, VINEGAR AND SUGAR

Serves 6

2 lb/1 kg of fresh blade shoulder of pork, 2 cloves of garlic, 4 sprigs of rosemary, 10 slices of pork fat, 1 cup of olive oil, 1 cup of red wine vinegar, salt and pepper to taste, 2 medium onions, sliced, 1 cup of red wine vinegar, 1 glass of white wine, 1 tbsp of sugar, 20 baby onions, 1 knob of butter

Method: 1. Lard the meat with pork fat, and spike with slivers of garlic and rosemary. Brush with oil and vinegar, and season with salt and pepper. Bind with kitchen twine and leave to marinate for 24 hours in a covered tureen.
2. The following day, brown the meat in oil. When it is nicely browned, add the sliced onions, vinegar, white wine and sugar. Simmer, covered, on a low heat for about 2 hours, adding stock if necessary. Remove the meat from the pan and strain the juices through a fine sieve.
3. Parboil the baby onions in water, vinegar and sugar, then add them to the pan juices, adding a knob of butter to enrich the sauce. Serve the meat in slices with the hot sauce.

Wines: Here, the intensely sweet flavor and richness of the pork contrasts deliciously with the other aromatic and slightly tangy ingredients. Choose a mature, full-bodied, dry, white wine, crisp but delicate, with a well-developed bouquet, such as Riviera Ligure di Ponente Pigato, Marzocco (Chardonnay) or Fiorano Semillon.

WALTER FERRETTO

SPICY PORK RISSOLES

Serves 8

11 oz/300 g of lean pork, ground, 11 oz/300 g of pig's liver, ground, 1 calf's brain, 2 eggs, 10 juniper berries, chopped, 2 tbsp of grated Parmesan cheese, pig's caul fat, salt, pepper, ground cinnamon and nutmeg to taste

Method: 1. Knead all the ingredients, except the caul fat, in a bowl and season with salt, pepper and nutmeg.
2. Divide the caul fat into 4½ inch/10 cm squares, place a tablespoonful of meat mixture on each. Wrap the caul fat around the meat and carefully transfer the rissoles to a roasting pan greased with oil.
3. Bake in a hot oven at 350°F/180°C for 15 minutes. Serve very hot with a tablespoonful of polenta and braised onions.

Wines: The subtle flavors of the main ingredients of this dish, pork, liver and brains, combined with the fairly intense taste of spices, calls for a delicate, dry, red wine with an intense fruity and floral bouquet, crisp, strong, fairly tannic and medium-bodied. Barbera del Monferato, Bardolino Classico or Pentro di Isernia would be good choices.

ALFONSO JACCARINO

PORK AND VEGETABLE SOUP

Serves 6

*11 oz/300 g of pork ribs, 4 oz/100 of pork fat, 11 oz/300 g of pig's snout and trotters, 7 oz/200 g of cotechino, 2 kg/4 lb of green vegetables
(broccoli, endive, borage, cabbage), 1 chili pepper, 7 tbsp/100 ml of dry white wine, salt*

Method: 1. Cut all the meat into small pieces and fry in the pork fat in a large handled pan. Add the wine and sauté for about 10 minutes.
2. Parboil the green vegetables and add them to the meat with 2 qt/2 l of cold water.
3. Chop the chili pepper and add to the soup with salt to taste. Bring to a boil and simmer slowly for 40 minutes. Leave to stand for 30 minutes before serving.

Wines: This rich, tasty soup should be accompanied by a delicate, crisp, young red wine with a rich, fruity bouquet, fairly alcoholic with a touch of tannin, such as Bramaterra, Colli Berici Tocai Rosso or Salina Rosso.

PORK RIND ROLLS WITH RAISINS AND PINE NUTS IN TOMATO SAUCE

Serves 4

1 lb 5 oz/600 of pork rind, ¼ cup/50 g of raisins, ¼ cup/50 g of pine nuts, ⅓ cup/20 g of parsley, 1½ oz/10 g of garlic, 3½ cups/300 ml of extra virgin olive oil, 2 cups/500 ml of tomato sauce, salt to taste

Method: 1. Clean the pork rind thoroughly, removing the bristle and fat.
2. Toast the pine nuts, chop the parsley and garlic. Fill the rind with raisins, pine nuts, parsley and garlic, then roll the rind around the filling and secure with kitchen twine.
3. Brown the rolls in a frying pan.
4. Bring the tomato sauce to the boil, then add the rolls to the sauce and simmer for about 1 hour. Serve very hot.

Wines: This is a harmonious and perfectly balanced blend of sweet, savory and aromatic flavors. To accompany it, choose a strong, full-bodied, mature red wine, with a rich bouquet of dried fruits and spices and a fair bit of tannin but still delicate and agreeable, such as Castel San Giorgio Riserva, Torgiano Rosso or Taurasi.

ERNST KNAM

PORK FILLET WITH PRUNES AND TEA SAUCE

Serves 4

14 oz/400 g of pork fillet, 12 pitted prunes, ⅔ cup/150 ml of hot tea, ⅔ cup/ 150 ml of brown veal stock, 2 oz/50 g of carrots, diced,
⅓ cup/20g of chopped parsley, salt and pepper

Method: 1. Soak the prunes in the hot tea for about 15 minutes.
2. Cut the pork fillet into eight equal pieces. Cut some of the prunes into thin strips. Make incisions in the meat and spike with the strips of prune.
3. Season the fillets with salt and pepper and fry them in a non-stick frying pan for 4-5 minutes, turning halfway. Remove from the heat and keep warm.
4. Add the tea to the stock and boil until reduced to a quarter.
5. Cook the remaining prunes and carrots in boiling salted water for about 1 minute, then drain and season with salt and pepper.
6. Arrange two pieces of meat on each plate, garnish with prunes and carrots. Pour over the sauce and garnish with a sprinkling of chopped parsley.

Wines: The sweet juiciness of the pork and prunes is counterbalanced by the more subtle flavors of the other ingredients. To complement this dish, serve a medium mature red wine, with an intense bouquet of red fruits and jam, strong and full-bodied but delicate with scarcely any tannin, such as Friuli-Annia Cabernet Franc Riserva, Colli Etruschi Viterbesi Canaiolo or Bivogno Rosso.

ERNST KNAM

PORK SHINBONES IN A MUSTARD AND APRICOT CRUST

Serves 4

4 pork shinbone, weighing a little over 1 lb/450 g each, 4 oz/100 g of onions, sliced, 4 oz/100 g of leeks, chopped, 4 oz/100 g of celery, chopped, 4 cloves of garlic, 7 tbsp of extra virgin olive oil, ¾ cup of brown veal stock, 5 tbsp/120 g of strong mustard, ½ cup/150 g of apricot jam, salt, freshly ground black pepper

Method: 1. Season the shinbones with salt and pepper and brown them in a preheated pan of olive oil.
2. Brown all the vegetables in a large roasting pan, then add the shinbones and roast in the oven at 475°F/220°C, moistening the meat with stock every 10 minutes.
3. After 40 minutes, mix the mustard and apricot jam and spread the mixture over the meat. Continue to cook for 1½-2 hours, until a crisp, golden crust forms.
4. Remove the meat from the pan and reduce the pan juices until they become syrupy.
5. Arrange the shinbones on a serving dish, covered with the sauce.

Wines: This mouth-watering recipe combines sweet, savory and aromatic flavors. Serve it with a crisp, delicate, young red wine, fairly alcoholic but with very little tannin, such as Terre de Franciacorta Rosso, Esino Rosso, or Alghero Cabernet.

ANGELO LANCELLOTTI

PORK RIND AND BEAN STEW

Serves 6

14 oz/400 g of pork rind, cleaned and singed over a flame to remove the bristles, and cut into 2 in/5 cm squares,
1½ cups/300 g of dried Sallugia beans (known in local dialect as "old lady's teeth") or borlotti beans (of which Sallugia are a variety), ½ onion, chopped,
2 tbsp of lard, 3 tbsp of passata (sieved tomatoes), salt and pepper to taste

Method: 1. Soak the beans in cold water overnight. The following day, bring an almost full pan of water to a boil, then add the beans and the rind, with a little salt, and boil until the beans are tender and the rind cooked *al dente*. (Provided the pig is not too old, this should take about 1 hour).
2. Now melt the lard in a frying pan and brown the onion, seasoned with salt and pepper. When the onion is nicely browned, transfer the rind and beans with a slotted spoon to the frying pan, adding a little of the stock.
3. Stand the pan on a heat diffuser over a low heat and simmer for about 1 hour to create a rind and bean stew. Serve very hot.

Wines: This succulent, fatty aromatic stew requires a delicate, young, dry, red wine, with an intense fruity, floral bouquet, crisp, strong, and medium-bodied with just a touch of tannin. Try Valtellina Rosso, Friuli-Grave Merlot or Guardiolo Rosso.

TORTELLINI IN "THREE MEAT" BROTH

Serves 4

For the pasta dough: 2½ cups/ 300 g of flour, 3 eggs
For the tortellini filling: 4 oz/100 g of pork loin, 4 oz/100 g of mortadella, 1 cup/100 g of Parma ham, 2 oz/50 g of sausage, 4 oz/100 g of Parmesan cheese,
1 egg, ¼ of a nutmeg, ½ tsp of fine salt, 2 tbsp of breadcrumbs
2 qt/2 l of broth, made with beef, chicken and a piece of ham (the legendary "three meat" broth, which used to be made only at Christmas)

Method: 1. Heap the flour onto a pastry board, make a well in the center, add the eggs, knead thoroughly then roll out. Cut the dough into 1½ in/3.5 cm squares.
2. Cut the pork, mortadella, Parma ham, and sausage into 1 in/2.5 cm pieces, then brown them gently in a frying pan for about 10 minutes. Turn the meat out onto a chopping board and, while still hot, chop it finely with a large knife.
3. Transfer to a bowl and add salt, grated nutmeg, grated Parmesan cheese, breadcrumbs and the egg. Mix thoroughly to produce a smooth filling. Place a teaspoonful of filling on each pasta square and fold over to create tortellini.
4. Bring the broth to a boil, add the tortellini and cook for 3-4 minutes, stirring very gently, until *al dente*.

Wines: To complement this classic combination of aromatic flavors, serve a strong, dry, well-structured, white wine, with a fruit and flower bouquet. Try something crisp and delicate, such as Tocai di San Martino della Battaglia, Collio Ribolla or Cacc'e Mmitte di Lucera.

PENNE WITH BLACK OLIVES, SHRIMPS AND PANCETTA

Serves 4

12 oz/350 g of penne, 2 oz/50 g of black olives, pitted and diced, 6 oz/160 g of shrimps, peeled and chopped (keep the heads), ½ cup/50 g of breadcrumbs, 1 tbsp of chopped parsley, 1 clove of garlic, chopped, 2 ripe tomatoes, diced, 1 glass of white wine, 1 glass of fish stock, 1 oz/30 g of smoked pancetta or streaky bacon, extra virgin olive oil, salt and pepper.

Method: 1. In a saucepan, brown the garlic in a little oil, then add the shrimps' heads, olives, pancetta and chopped shrimps. Cook over a low heat for a few minutes, then add the tomatoes, wine and fish stock. Simmer for a few minutes, then remove the shrimps' heads, and keep the sauce warm.

2. Cook the penne in plenty of salted water, until *al dente*.

3. Meanwhile, gently toast the breadcrumbs in a frying pan with a little oil until golden.

4. Drain the pasta thoroughly and stir in the breadcrumbs, followed by the sauce. Stir in the parsley and sauté over a low heat for 1 minute, then drizzle with a little olive oil and sprinkle with pepper (or chili powder). Serve immediately.

Wines: The curious mixture of shrimps, pork and black olives produces a balanced blend of intense flavors. With this dish, serve a very crisp, dry white wine, strong and full-bodied, but delicate, with a fresh, fruity, floral bouquet, such as Verdicchio dei Castelli di Jesi, Vernaccia di San Gimigniano or Menfi Grecanico.

VALENTINO MARCATTILII

SADDLE OF PORK IN MILK

Serves 4

2 lb/1 kg of boned saddle of pork, 2 ribs of celery, 1 large onion, 2 carrots, 8 juniper berries, 2 cloves, 6 cups/1.5 l of milk, ½ cup/100 g of butter, 1 glass of white wine, salt and freshly ground black pepper

Method: 1. Truss the meat and season with salt and pepper.

2. Wash and dice the celery, carrots and onion. Crush the juniper berries and cloves. Soften the chopped vegetables in an ovenproof stewpot with the butter, then stir in the cloves and juniper berries, before adding the meat. Cook over a low heat for 30 minutes, moistening the meat with white wine. When most of the wine has been absorbed, add the milk. Cover the pot and cook in the oven at 325°F/160°C for 1 hour.

3. Remove from the oven, take the meat from the pot, carve and arrange the slices on a serving dish, and keep warm. Strain the pan juices through a sieve, season with salt and pepper, and pour some of the sauce over the meat. Serve the remainder in a sauceboat. Serve the pork with sweet-and-sour baby onions.

Wines: To complement the subtle sweetness of the pork and milk, combined with the aromatic flavors of the other ingredients, choose a strong, young, dry red wine, with an intense bouquet of fruit and flowers, crisp and delicate with minimal tannin. Pignolo del Fruili, Cagnina di Romagna or Santa Margherita di Belice Nero d'Avola would be ideal.

SERGIO MEI

PORK RIND SALAMI WITH VEGETABLE AND HERB STUFFING

Serves 4

For the salami: 7 oz/200 g of pork rind, 4 oz/100 g of pork shoulder, 2 oz/50 g of duck liver, 2 cloves of garlic, 1 tsp/5 g of chopped shallots, grated lemon zest, 4 tbsp/50 ml white wine, 4 tsp/20 ml red wine vinegar, ½ oz/10 g sun-dried tomatoes, salt and pepper
For the herb stuffing: 1 oz/30 g white bread, crusts removed, 1 clove of garlic, 1 shallot, 1 oz/30 g parsley, several sprigs of rosemary, a small sprig of mint, 1 tsp of pine nuts, 4 tbsp of pork fat. For the trimmings: 7 oz/200 g of cabbage, 1 oz/50 g of fennel, ½ oz/20 g of honey fungus caps or other wild mushrooms, 2 tsp of raisins, 2 tsp of black olives, 2 oz/50 g of leeks, ½ oz/15 g of shallots, 2 tsp of extra virgin olive oil. For the garnish: a sprig of savory, 8 raspberries. For the stock: 1¾ lb/800 g of pork bones, 2 oz/50 g of celery, chopped, 1 oz/30 g of carrots, chopped, bouquet garni, 6 oz/150 g of tomatoes, 2 tsp of honey, 2 cups of vernaccia or other dry white wine, 1 tsp of coarse salt, 4 tsp of extra virgin olive oil, 2 qt/2 l of water, 3½ tbsp of butter, 10 cloves of garlic, 2 oz/50 g of shallots, 4 tbsp of red wine vinegar, pepper

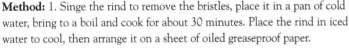

Method: 1. Singe the rind to remove the bristles, place it in a pan of cold water, bring to a boil and cook for about 30 minutes. Place the rind in iced water to cool, then arrange it on a sheet of oiled greaseproof paper.
2. To make the meat stuffing, mince all the ingredients, apart from the shallots, wine and vinegar, which should be boiled in an uncovered saucepan, until reduced by at least half. To make the herb stuffing, grind the pine nuts, bread and pork fat in a blender, add the other ingredients and process until smooth.
3. Spread the meat stuffing over the rind, then add the herb stuffing in middle of the rind. Roll up the rind and bind it with kitchen twine.
4. Chop the bones into small pieces, and brown them in butter and coarse salt in a large pot with the celery, carrot and bouquet garni. Add the rolled salami, brown briefly, and place in the oven at 350°F/180°C. Pour the white wine over the salami and cook for about 25 minutes. When the salami is cooked, remove it from the pot and leave it to rest for five minutes.
5. Meanwhile, slice the leeks lengthwise into long strips, wash thoroughly and blanch them for about 1 minute in boiling salted water. Untie the salami and carve it into 1 in/2.5 cm slices, bind each slice with a strip of blanched leek, and keep warm. Add the water to the pan juices and boil over a high heat until reduced by half. Strain the liquid through a fine sieve into another saucepan and simmer over a low heat until it becomes syrupy, seasoning with honey, the reduced vinegar, salt, pepper and oil.
6. Shred the cabbage and fennel into julienne strips and sauté with the mushrooms, shallots, raisins, pine nuts and olives.
7. To serve, arrange a bed of cabbage and fennel on the plate, lay a slice of salami on top, and arrange the raspberries and sautéed mushrooms round the edge. Sprinkle with the sauce, mushroom mixture, and sprigs of savory.

Wines: The variety of ingredients in this dish creates a delicious combination of sweet and savory, richness and lightness. Serve with a mature, very acidic, red wine, with a well-developed bouquet of fruit, flowers and spices, and plenty of alcohol, such as Dolcetto de Ovada, Colli di Scandiano, Canossa Malbo Gentile or Settesoli Rosso.

FABIO MOMOLO

HERBY PIG'S LIVER PARCELS

Serves 6

A generous pound/500 g of pig's liver, 7 oz/200 g of pig's caul fat, 4 oz/100 g of pork fat, chopped into small pieces, 1 onion, 8 rosemary leaves, coarsely chopped, 8 sage leaves, 1 cup/100 g of fresh parsley, coarsely chopped, ¾ cup/200 ml of dry white wine, salt and pepper

Method: 1. Cut the liver into neat pieces and wrap each piece in caul fat with a sage leaf.
2. Cut the onion into thin rings, melt the pork fat in a frying pan and fry the onion rings for a few minutes, then add the chopped rosemary and parsley.
3. Add the liver parcels and cook over a moderate heat for at least 20 minutes, adding salt, pepper and white wine.
4. Serve on a slice of toasted polenta with stewed beans.

Wines: This is a complex blend of sweet and savory, which also combines aromatic flavors with a touch of sharpness and slight bitterness. To accompany it, choose a well-structured, full-bodied, young red wine with a rich, fruity aroma, delicate and discreetly tannic, and pleasantly warming. Freisa delle Langhe, Chianti di Montalbano or Gioia del Colle Primitivo would be good choices.

AIMO & NADIA MORONI

STUFFED LEG OF PORK SPIKED WITH PORK FAT ROULADES AND GRAPE SAUCE

Serves 12

1 leg of pork weighing about 15 lb/7 g, 1 lb/500 g of muscatel grapes, 2 oz/50 g of pork fat, ¾ cup/200 ml of meat stock, ¾ cup/200 ml of white wine, 1 carrot, 1 onion, 1 rib of celery, 1 sprig of rosemary, 2 sage leaves, 1 clove of garlic, salt and pepper

Method: 1. Singe the leg of pork over a flame to remove any bristles, then wash and dry with a clean cloth. Using a small skinning knife, remove about ⅔ of the rind from the leg and boil it in salted water for about 1 hour. 2. Meanwhile, bone and trim the leg, reserving the trimmings. With a large knife, chop the pork fat, pork trimmings, rosemary, sage and garlic and mix together, seasoning with salt and pepper. Sprinkle the inside of the leg generously with salt and pepper, then fill with ⅔ of the meat mixture. 3. When the rind is cooked, carefully remove all the fat, and cut it into rectangles. Spread the remaining mixture over the inner surface of the rectangles, then roll them up tightly and tie with kitchen twine. Brown them in a frying pan over a moderate heat for a few minutes, without adding fat, then remove the twine. 4. Make incisions in the leg of pork deep enough to hold the rind roulades. Roast the meat in the oven at 350°F/180°C for at least 3 hours. Moisten the

meat with wine and, when it has all evaporated, add stock, as required. Wash the grapes and remove the stalks, heat them in a saucepan for a few minutes, then grind in a blender and strain the juice through a fine sieve. 5. When the meat is cooked, remove it from the pan and keep warm. Add the sieved grapes to the pan juices, skim off the fat and reduce the sauce over a moderate heat, then strain it through a sieve. Serve the meat in slices, with the grape sauce.

Wines: The acidity of the grapes provides a perfect contrast to the sweet, rich, fatty meat. As an accompaniment, serve a well-structured, medium-mature red wine, crisp, strong but delicate, and slightly tannic, such as Monfortino, Bolgheri Sassicaia or Colle Picchioni Rosso.

| AIMO & NADIA MORONI |

PORK ROULADES WITH SWEET ONION AND POTATO CROQUETTES

Serves 4

1½ lb/700 g of loin of pork, cut into four pieces, 7 oz/200 g of pig's caul fat, 4 oz/100 g of pork fat, shredded, 2 lb/1 kg of lard, 1 tsp of chopped sage and rosemary, 1 clove of garlic, chopped, 1 pinch of nutmeg
For the stock: a generous pound/500 g of pork bones and rind, in pieces, 3 tbsp each of chopped onions, carrots, and leeks, 2 cups of beef stock, 1 glass of medium dry white wine, 1 bay leaf, 1 sprig of thyme, 2 tbs of butter, 4 tbsp of extra virgin olive oil
For the purée: 1¼ lb/600 g of onions, peeled and chopped, 1 large boiled potato, 2 tbsp of acacia honey, ½ glass of beef stock, 3½ tbsp/50 g of butter, toasted pine nuts, salt and pepper

Method: 1. Brown the onions in the butter, add the stock and cook over a medium heat for about 10 minutes. Add the peeled and sliced potato and the honey and cook for a further 10 minutes. Pass the mixture through a vegetable mill, season with salt, and keep warm.
2. Parboil the bones for 5 minutes and singe the rind over a flame to remove any bristles. To make the sauce, brown the chopped onions, leeks and carrots in a frying pan with olive oil, bay leaf and thyme. Add the bones and rind, and when these are nicely browned add the wine. When the wine has evaporated, add the stock and simmer for about 45 minutes, then strain and skim off the fat.
3. Rinse and dry the caul fat, then spread it out and divide it into four pieces. Mix the pork fat, sage, rosemary, garlic, nutmeg, salt and pepper.
4. Make three incisions in each piece of meat and stuff with the mixture, then wrap in the caul fat, and secure each piece with three toothpicks.

5. Melt the lard in a frying pan and cook the roulades for 45 minutes, then remove from the heat and keep warm.
6. Reheat the sauce, adding the butter at the end.
7. Meanwhile, remove the caul fat from the meat, cut each piece of meat into ½ in/2.5 cm slices, and arrange on very hot plates, covered with sauce. Shape the purée into croquettes and garnish with toasted pine nuts.

Wines: This dish provides a contrast between sweet and savory tastes combined with the rich, aromatic flavors of the various ingredients. Serve with a strong, full-bodied red wine, dry, crisp, delicate and not too young, with a rich bouquet of fruit, flowers and herbs, such as Carema, Colli Berici Cabernet or Chianti di Montalbano.

ARNEO & DARIO NIZZOLI

PIG'S LIVER IN TOMATO SAUCE

Serves 8

3 lbs/1.5 kg of pig's liver, 1 lb/500 g of pig's caul fat, 3½ tbsp/50 g of butter, 1 tbsp of tomato purée, 1 glass of white wine, 1 onion, salt and pepper

Method: 1. Cut the caul fat into pieces and the onion into rings and brown in butter in a frying pan, stirring frequently.
2. Meanwhile, cut the liver into thin strips, and dissolve the tomato purée in the white wine. Add the liver to the pan and, when it is half cooked, add the wine and tomato purée, stir thoroughly and add the seasonings. The liver is cooked when it no longer gives off any blood.
3. Serve with boiled polenta.

Wines: In this very simple dish, the sweetness of the liver and caul fat is counterbalanced by the piquancy of the other ingredients. Choose a very crisp, young, dry white wine, medium-bodied and fairly delicate with an intense bouquet of fruit and flowers, such as Venegazzù Pinot Bianco, Ansonica Costa dell'Argentario or Vallarom Bianco.

RICE WITH PIG'S LUNG

Serves 8

1½ cups/320 g of risotto rice, 1½ lb/600 g of pig's lung, finely diced, 1 large onion, 1 rib of celery, 1 tbsp of tomato purée, 1 glass of white wine, mixed spices, salt, pepper and cinnamon, 2 oz/60 g of pork fat, 2 tsp/10 g of butter, 2 tbsp of grated Parmesan cheese, 2 qt/2 l stock or water.

Method: 1. Finely chop the onion and celery and brown them in a frying pan with a knob of butter and the pork fat. As soon as they start to take color, add the lung, and cook briefly before adding the wine. Allow the wine to evaporate and then add the tomato purée and the seasonings and simmer for a few minutes.

2. Transfer the mixture to a saucepan, simmer for 15 minutes, then add boiling water or stock, season with salt and simmer over a low heat for at least 15 minutes longer. Then add the rice and continue to cook until the rice is tender.

3. Serve very hot, sprinkled with freshly ground black pepper and, if you wish, some grated cheese.

Wines: This dish in the popular tradition is rich, sweet and aromatic. Serve with a dry but delicate white wine, crisp and strong, with a fruity, floral bouquet, such as Riviera del Garda Bresciano Groppello, Colli di Conegliano or San Vito di Luzzi.

DAVIDE OLDANI

STUFFED PIG'S TROTTER

Serves 6

2 pig's trotters, ¾ cup/200 ml of white wine, 1 bouquet garni, 7 tbs of olive oil, 3 qt/3 l veal stock
For the stuffing: 11 oz/300 g of pork loin, 5 oz/150 g of salted pork fat, 4 oz/100 g foie gras, 2 oz/40 g of aromatic herbs (chives, chervil, parsley), 1 cup of milk,
4 slices of white bread, crusts removed, 2 eggs, 2 oz/40 g of chopped shallots, 4 oz/100 g of porcini mushrooms, 1 lb/500 g of pig's caul fat,
port and brandy to taste, salt and pepper

Method: 1. Clean the trotters thoroughly and brown them in a saucepan with olive oil. Pour off the fat and deglaze the pan with white wine. When the wine has evaporated, add enough veal stock to cover the meat, and the bouquet garni, then bake in the oven for 2 hours at 275°F/130°C. Allow to cool, strain the pan juices, and bone the trotters, taking care not to tear the meat.
2. Meanwhile, prepare the stuffing. Soak the bread in milk, squeeze it dry, then put it through the mincer with the meat and pork fat. Brown the chopped shallots in a little oil. Cut the *foie gras* into small pieces and brown it a separate frying pan. Clean and thinly slice the porcini mushrooms and sauté them in a little oil in a separate pan.
3. Transfer the meat mixture to a bowl, adding the chopped herbs, shallots, egg, *foie gras*, brandy, port, porcini, salt and pepper. Blend thoroughly with a wooden spatula. Stuff the trotters and wrap them first in the caul fat and then in foil, put them in a pot in a double boiler and cook in the oven at 350°F/180°C for about 1-1½ hours.
4. Cut into 1 in/2.5 cm slices and serve with other roast meats.

Wines: This ancient recipe, adapted to suit our modern tastes, is wonderfully succulent and aromatic. It calls for a strong, full-bodied young red wine, delicate and crisp, with a fruity bouquet, such as Fara, Breganze Cabernet or Pollino.

HEADCHEESE TERRINE

Serves 6

2 lbs/1 kg of pig's head, boned and trussed, 2 carrots, 1 onion, 1 rib of celery, 1 bay leaf, 1 bunch of parsley, 4 qt/4 l of water, coarse salt to taste
For the garnish: 2 shallots, 2 carrots, 3 tbsp of chopped herbs (chervil, parsley, chives, tarragon), 7 oz/200 g of porcini mushrooms,
2 tbsp of extra virgin olive oil, salt and pepper

Method: 1. Make vegetable stock with the ingredients listed above, and boil the pig's head in the stock for 2½ hours.
2. Meanwhile, prepare the garnish. Chop the shallots and soften them in olive oil. Dice the carrots and blanch them in boiling salted water. Clean and chop the mushrooms and sauté them in a frying pan with olive oil.
3. Allow the pig's head to cool, then untie it, chop the meat coarsely and place it in a bowl and season with salt and pepper. Stir in the shallots, chopped herbs, carrots and mushrooms. Transfer to a terrine greased with oil. Stand the terrine in a double boiler and cook in the oven at 350°F/180°C for 1½-2 hours.

Note: This terrine can be served cold, thinly sliced and dressed with a vinaigrette made with balsamic vinegar, or sliced, dipped in breadcrumbs, fried, and served with roast potatoes and black truffles.

Wines: This appetizing terrine strikes a delicious balance between the richness and sweetness of the meat and the subtle, aromatic flavor of the herbs and other ingredients. Serve with a young, very crisp, red wine, full-bodied with plenty of tannin, such as Colline Lucchesi rosso, Colli Amerini rosso, or Primitivo di manduria secco.

PORK RIND WITH CELERY HEARTS AND POTATOES

Serves 10

1½ lb/600 g of pork rind, 1 onion, 1 rib of celery, 1 carrot, 2½ tbsp/40 ml of good olive oil, 2 bay leaves, 1 sprig of sage, a few sprigs of rosemary,
4 cloves of garlic, crushed, freshly ground black pepper, chili powder, 3 cups of peeled tomatoes, 5 celery hearts, sliced, 15 small potatoes,
1 ladleful of stock or water, salt

Method: 1. Cut the pork rind into finger-length strips, then plunge them in boiling water and boil for 1½ hours. Drain and rinse under running water.
2. Finely chop the onion, celery and carrot and fry them in the olive oil. When the vegetables are nicely browned, add the bay leaves, a few sage leaves, a few sprigs of rosemary and the cloves of garlic. Season with pepper and chili powder (the more the better).
3. Now add the pork rind, the peeled tomatoes, the sliced celery hearts, and the potatoes. Add a ladleful of stock or water, and simmer slowly for about 1 hour. Add salt to taste. Serve with some good bread, an absolute necessity with this dish.

Wines: This simple but tasty winter dish combines the rich succulence of the pork rind with the aromatic flavors of the other ingredients. Serve with a well-structured, young, red wine with plenty of alcohol, crisp and delicate, with a touch of tannin and an intense fruity, floral bouquet. Try Breganze Pinot Nero, Rosso di Montalcino, or Cerasuolo di Vittoria.

CHARCOAL GRILLED PORK CHOPS

Serves 10

6½ lb/3 kg of pork spare ribs in one piece, salt, pepper, the juice of 6-7 lemons, 2 bunches of sage, 2 sprigs of rosemary, the zest of 1 lemon

Method: 1. Choose the spare ribs of a good, fat pig, divide into chops then marinate them in plenty of lemon juice for 6-8 hours.
2. Make sure the charcoal embers are very hot, then season the chops generously with salt and pepper. The chops will cook quickly after being marinated in lemon juice, which also lends delicious flavor to the fatty meat. When they are almost ready, you can, if you wish, season the chops with finely chopped sage, rosemary and grated lemon zest.

Wines: The best complement to this simple but delicious and aromatic recipe would be a mature, well-structured red wine, crisp, delicate, and slightly tannic, with a well-developed bouquet of fruit, flowers and spices. Try Collio Merlot, Sant'Agata dei Goti Piedirosso or Castel del Monte Pinot Nero.

CLAUDIO PRANDI

RIBS AND TROTTERS WITH SAUERKRAUT

Serves 4

6 oz/150 g of pig's cheek, 2 fresh pig's trotters, 1½ lb/1.2 kg of pork rib chops, 1½ lb/600 g of pickled sauerkraut, 1 bay leaf, 4 whole peppercorns, 1 clove of garlic,
a little grated horseradish, 1 onion, chopped, 1 Reinette or other tart, crisp apple, chopped, 1 glass of dry white wine, 1 qt/1 l of chicken stock,
3 tbsp of extra virgin olive oil, salt and pepper to taste

Method: 1. Boil the trotters in plenty of water for ½ hour, then drain and remove the bones. Cut the rib chops into pieces, boil in salted water for 2 minutes, then drain and set aside.
2. Wash and dry the sauerkraut. Cut the pig's cheek into small pieces, then fry it with the chopped onion and whole clove of garlic in a saucepan with olive oil. When the ingredients begin to take color, discard the garlic, and add the sauerkraut. Allow the liquid to evaporate for 10 minutes, then add the white wine, trotters and ribs. Cover and cook over a low heat for 30 minutes, then add the horseradish, apple, bay leaf and peppercorns. Moisten with a little stock and continue to cook, covered, for at least 1 hour.
3. Check to see that the meat it cooked and adjust the seasoning. Serve with polenta.

Wines: This recipe pleasantly unites the sweetness and succulence of the meat with the aromatic and slightly sharp flavor of the sauerkraut, horseradish and other ingredients. As an accompaniment, choose a delicate, young red wine, crisp and full-bodied with only a touch of tannin and an intense, fruity, floral bouquet, such as Carso Merlot, Montescudaio Rosso or Lacrima di Morro.

PORK AND BEAN HOTPOT

Serves 4

1 cup/200 g of dried Saluggia or borlotti beans, soaked in cold water overnight, 11 oz/300 g of pork rib chops, 2 pig's trotters, 4 oz/100 g of pork rind, rosemary, 1 bay leaf, sage, salt and pepper, 2 onions, 2 potatoes, 1 carrot, 2 ribs of celery, 1 bouquet garni, 1 glass of red wine

Method: 1. Drain the beans and place them in a pan of cold, salted water. Bring to a boil and cook for 1-1½ hours until *al dente*, then drain.
2. Divide the pork rind into four pieces and mince it with the rosemary, bay leaf, sage, salt and pepper. Brown the meat and rind in a frying pan with very little oil over a high
3. Finely chop the onions, carrot and celery and brown them in a little oil. Wash and peel the potatoes.
4. Transfer the meat, vegetables, whole potatoes and beans to a heatproof earthenware pot. Turn up the heat, add the red wine and cook until it evaporates. Add the bouquet garni and enough cold water to cover all the ingredients. Cook in the oven at 350°F/180°C for about 2 hours.
5. When the hotpot is ready, mash the potatoes and return them to the pot. Serve very hot.

Wines: This rich hotpot, in which the flavors of the rib chops and rind predominate, is best complemented by a young, strong and medium-bodied, possibly slightly sparkling, red wine, which is crisp, delicate and slightly tannic, with a fragrant bouquet of fruit and flowers. Oltrepò Pavese Buttafuoco, Breganze Rosso or Santa Margherita di Belice Rosso would be ideal.

ADRIANO PRESBITERO

POLENTA, TURNIPS AND SAUSAGE

Serves 4

2 lb/1 kg of turnips, 1¼ cups/200 g of butter, 1 heaped tbsp of sugar, 11 oz/600 g of very fresh pork sausage, 1 sprig of rosemary, salt and pepper

Method: 1. Peel the turnips and cut them into small pieces. Brown the butter with the sugar in a large frying pan, then add the turnip. Cover and cook slowly over a low heat.
2. In a separate pan, brown the sausage in a little oil, adding a sprig of rosemary.
3. Season the turnips lightly with salt and pepper, drain off the fat from the sausage before adding it to the turnips. Cook over a high heat for a few minutes.
4. Serve very hot with fine-grain polenta.

Wines: This rich, juicy and discreetly aromatic dish should be accompanied by a strong, medium-bodied, possible lightly sparkling, young red wine, fairly delicate and crisp, with just a touch of tannin and a fruity bouquet. Choose a Monferrato Barbera, Colli Martani Sangiovese, or San Severo Rosso.

ROMANO RESEN

CRACKLING TART WITH RASPBERRY JAM

For a 12 in/26cm cake tin

9 oz/250 g of fresh pork crackling, finely chopped then passed through the mincer, 2 cups/250 g of flour, ½ cup/150 g of sugar, 1 egg, a pinch of cinnamon,
2 cloves, crushed, grated lemon and orange zest, ½ tbsp of baking powder
For the raspberry jam: 6 cups/1 kg fresh raspberries, sugar, zest of 1 lemon, raspberry schnapps
To serve: raspberries, lemon juice, confectioner's sugar, unsweetened whipped cream

Method: 1. Prepare the pastry, blending the egg and sugar, crackling, spices, grated zest, flour and baking powder. Wrap the dough in cellophane film and refrigerate for at least 2 hours before using.
2. To prepare the raspberry jam, crush 6 cups/1 kg of fresh raspberries with a fork in a heavy-bottomed pan and let them dry out over a low heat. Stir constantly with a wooden spoon to ensure that the liquid does not dry out completely. Leave to cool. Weigh the remaining raspberry pulp, then add an equal quantity of sugar, the zest of one lemon and a shot glass of raspberry schnapps. Place over a low heat and continue to cook, stirring constantly, until the jam thickens.
3. Roll out the pastry. Line a buttered cake tin (preferably also lined with greaseproof paper) with a layer of pastry, with a raised edge at least 1-1½ in/2.5-3 cm high. Spread the pastry shell with raspberry jam, and

cover it with a lattice made with strips of the remaining pastry. Bake in the oven for 20-25 minutes at 350°F/180°C.
4. Serve warm or cold with a sauce of raspberries whisked with lemon juice and confectioner's sugar, and a spoonful of unsweetened whipped cream.

Wines: This rather odd combination of ingredients is rich, sweet, spicy and slightly sharp. It should be served with a medium dry or sweet white wine, not too young, with a rich, well-developed bouquet of yellow fruits, flowers and spices, still pleasantly acidic, but delicate, with lots of alcohol and plenty of body. Try Gambellara Vinsanto, Muffato della Sala or Semidano di San Vero Milis.

ROMANO RESEN

PORK RIB SPIKED WITH STURGEON WITH FRIED SAGE, POTATOES AND RED ONIONS

Serves 4

2 double pork ribs weighing about 1 lb/500 g each, with longer bones than usual, 4 oz/100 g of thinly sliced sturgeon, salt, pepper, chives, the juice of 1 lemon, 4 tbsp/50 ml of extra virgin olive oil, 4 tbsp of butter
For the sauce: 7 oz/200 g of fresh sturgeon, diced, 2 oz/50 g of smoked sturgeon, 4 tsp of extra virgin olive oil, 4 tbsp/20 g of butter, 7 tbsp/100 ml of veal stock, 7 tbsp/100 ml of broth, 7 tbsp/100 ml of mature red wine vinegar
For the garnish: 8 sage leaves, anchovy paste, 1 tbsp of flour, 7 tbsp/100 ml of beer, the yolk of 1 egg, ice, oil for frying, 14 oz/400 g of small, new potatoes, 7 oz/200 g of shallots, 4 tbsp/20 g of butter

Method: 1. Trim the meat, without removing the fat. Dip the thinly sliced sturgeon in a mixture of salt, pepper and chopped chives. Leave to macerate in olive oil and lemon juice for at least 2 hours.
2. To prepare the sauce, brown the fresh and smoked sturgeon in olive oil and butter, moistening the fish with equal quantities of veal stock, broth and red wine vinegar. Cook until slightly reduced, then purée in a blender.
3. Season the meat with salt, and fry in a large, non-stick frying pan with the oil and butter until browned all round. Skim off the fat from the pan juices and stir in the sauce.
4. Transfer the meat and sauce to the oven and roast at 350°F/180°C for about 1½ hours, adding more stock if the meat becomes too dry. Remove the meat from the oven and keep warm.
5. Finish cooking the sauce, adding a knob of butter and chopped parsley.
6. To prepare the garnish, make a batter with the flour, beer, egg yolk and a little crushed ice. Spread four of the sage leaves with anchovy paste, then cover with the remaining leaves to make "sandwiches". Dip the leaves in batter and deep fry in olive oil over a medium heat, then drain on kitchen paper. Steam the potatoes and shallots, then brown them in hot butter with salt. Arrange the meat on a serving dish. Carve the joint crosswise into fairly thin slices, and garnish each portion with a sage leaf. Cover with sauce and serve the vegetables and remaining sage leaves separately.

Wines: The curious but pleasant combination of meat and fish is sweet, tasty and subtly aromatic. Serve with a very delicate, young red wine, crisp, with a fruity bouquet, not too much tannin, and plenty of alcohol, such as Verduno Pelaverga, Colli Euganei Merlot or Rosso di Andria.

SWEET AND SOUR LOIN OF PORK WITH ONIONS

Serves 4

A generous lb/500 g of loin of pork, 5 onions, sliced, 8 tbsp/80g of butter, 2 tbsp/30g of sugar, 1 tbsp of flour, 7 tbsp/100 ml of red wine vinegar, 7 tbsp/100 ml of red wine, 2 cups/500 ml of meat stock, 1 bay leaf, salt and pepper

Method: 1. Melt the butter in a stewpot and, when it froths, add the sliced onion and fry until well browned. Add the sugar, vinegar and the bay leaf and leave over a medium heat to reduce.
2. Cut the meat into small pieces, coat with flour, then brown it in a separate pan with a little oil and a knob of butter.
3. Add the meat to the onions and add the wine. When the wine has evaporated, add the stock and season with salt and pepper. Simmer over a low heat until the sauce is fully reduced.

Wines: The beautifully balanced contrast between sweet and sour in this dish calls for a strong, full-bodied, but very delicate, fairly crisp, red wine with very little tannin and an intense fruity, floral bouquet. Try Valcalepio Rosso, Colli Bolognesi Cabernet Sauvignon or Cilento Rosso.

ELIA RIZZO

PORK RIND AND BEAN RISOTTO

Serves 4

*1½ cups/280 g of vialone nano or other risotto rice, ½ cup/100 g of dried borlotti beans, soaked in cold water overnight, 4 oz/100 g of pork rind,
½ cup/50 g of grated Parmesan cheese, 1 oz/30 g of chopped onion, 2 tbsp of extra virgin olive oil, 9 tbsp of butter, 4 cups/500 ml of vegetable stock,
chopped parsley, black pepper, salt*

Method: 1. Rinse the beans and place them in a pan of cold, salted water. Bring to a boil, cook for 1-1½ hours until *al dente*, then drain thoroughly.
2. Boil the pork rind in salted water for about 30 minutes, then drain thoroughly and cut into strips about ½ in/1 cm wide and 1½ in/3cm long.
3. Brown the chopped onion in 4 tbsp butter and olive oil. Add a ladleful of stock, and then, after about 20 minutes, add the beans and simmer for a further 10 minutes.
4. Purée about ⅓ of the beans. Add the rind to the rest of the ingredients and simmer for another 5 minutes. Check the seasoning and stir in the puréed beans.
5. Toast the rice in 4 tbsp of butter, then add the rest of the boiling stock, and continue to cook until the rice is tender and fairly dry.

Wines: To complement this rich combination of the intense sweetness of the rice and beans and the salty, aromatic flavors of the other ingredients, choose a strong, full-bodied, young, red wine, delicate and crisp with an intense bouquet of fruit and flowers and a touch of tannin. Oltrepò Pavese Rosso, Lison Pramaggiore Merlot or Falerno del Massico Primitivo would be ideal.

6. Stir in the beans and rind, grated cheese, chopped parsley, a sprinkling of freshly ground black pepper, and the rest of the butter, and serve.

PORK FILLET IN A CRUST

Serves 10

2 lb/1 kg of puff pastry, 1¾ lb/800 g of pork fillet, 7 oz/200 g of prosciutto crudo, 7 oz/200 g of lean minced pork, the yolks of 2 eggs, 1 shot of brandy, 4 tbsp of butter, salt and pepper to taste

Method: 1. Mix the minced pork with the egg, brandy, salt and pepper.
2. Butter a rectangular baking sheet and line it with half the puff pastry. On top, arrange alternating layers of thinly sliced pork fillet, *prosciutto crudo* and minced pork. Cover with the rest of the pastry.
3. Bake in the oven at 350°F/180°C for about 2 hours, then leave to cool. Serve cold, cut into thick slices.

Wines: This classic recipe blends the sweet juiciness of the pork fillet with the more subtle, aromatic flavors of the other ingredients. As an accompaniment, serve a mature, dry, red wine, strong, full-bodied, and fairly tannic, with a well-developed fruity, floral bouquet, such as Gattinara, Valtellina Sfurzat or Brunello di Montalcino.

ANDREA ROSSI

SMOKED LEG OF PORK WITH TRUFFLES

Serves 10

4½ lb/2 kg of smoked leg of pork, 11 oz/300 g of calf's or goose liver pâté, 4 oz/100 g of cooked ham in one slice, 7 tbsp/50 g of pistachios, 1 small black truffle, 1 shot glass of Cognac, salt and pepper to taste

Method: 1. Bone the leg of pork and spread it out to make room for the stuffing.
2. Dice the ham and cut the truffle into thin slivers. Mix the pâté with the ham, pistachios, truffle and Cognac.
3. Fill the leg with the stuffing, then sew it back together, wrap in baking parchment or foil and bind very tightly with kitchen twine. Boil for 2 hours, then leave to cool.
4. Serve cut into ¼ in/5 mm slices, either as it is or with gelatine.

Wines: To complement the richness of the meat and the subtle taste and texture of the truffles and pistachios, choose a crisp, young, red wine, strong, smooth and slightly tannic with an intense fruity, floral bouquet. Good choices include Breganze Pinot Nero, Rosso di Montalcino or Cerasualo di Vittoria.

ROMANO ROSSI

SALAMI AND CHEESE RAVIOLI WITH COCOA STRIPES

Serves 6

For the pasta dough: 3¼ cups/400 g of flour, 1 tsp of unsweetened cocoa powder, 1 tsp of vegetable oil, a pinch of salt
For the filling: 7 oz/200 g of ready cooked salamina (meaty sausage from Ferrara), 1 medium potato, boiled and peeled, 4 oz/100 g squacquarone, stracchino, taleggio
or similar creamy cheese, ½ cup/80 g of grated Parmesan cheese, 1 egg, salt and pepper, ⅓ cup/80 g of butter for frying

Method: 1. Mix all the ingredients for the filling in a blender to create a smooth, creamy paste.
2. Knead together all the ingredients for the dough, apart from the cocoa, then take a quarter of the dough and blend in the cocoa. Roll the plain and cocoa-flavored dough in separate sheets.
3. Cut the cocoa dough into narrow strips. Arrange the strips diagonally on top of the yellow dough and press down firmly. Then cut the dough into 2½-3 in/5-6 cm squares. Place the filling in the middle of each square, fold in half to make triangles, then press around the edges with the prongs of a fork to seal them tightly.
4. Cook these ravioli in plenty of boiling, salted water. As soon as they float to the surface, remove them from the water with a slotted spoon.
5. Melt the butter in a frying pan and toss the ravioli in the hot butter before serving.

Wines: The unusual use of cocoa in these delicious ravioli makes them even more appetizing. To accompany them, serve a strong, smooth, crisp, red wine with no more than a hint of tannin and an intense bouquet of fruit and flowers, such as Roero Rosso, Garda Orientale Marzemino or Verbicaro Rosso.

SMOKED SNAILS WITH SPECK, NECK OF PORK AND SMOKED SCAMORZA CHEESE

Serves 4

24 snails, 7 tbsp/100 g of butter, 1 slice of neck of pork weighing about 3½ oz/100 g, grilled, 2 oz/50 g of speck (cured ham),
2 oz/50 g of smoked scamorza (buffalo cheese), 1 clove of garlic, 1 glass of dry white wine, 1 rib of celery, 1 shallot,
1 small carrot, salt and pepper to taste.

Method: 1. Coarsely chop the celery, shallot and carrot. Wash the snails thoroughly and place them in a saucepan with the chopped vegetables, wine and plenty of cold water. Cook for 4-5 minutes, adding more water from time to time, if necessary. Do not add salt and pepper until the snails are cooked. Leave to cool in the stock.
2. Grind the butter, pork, speck, scamorza, and garlic in a blender. Season with salt and pepper, and blend until the mixture is smooth and even.
3. When the snails are cold, remove them from their shells. Then put a little meat mixture in each shell, replace the snail, and finish with a little more meat mixture, smoothing the surface.
4. Bake in the oven at 400°F/200°C for 20-25 minutes, and serve very hot.

Wines: This rich, savory and aromatic dish should be accompanied by a robust, crisp, nicely tannic wine, smooth and well-structured with an intense fruit and flower bouquet. Try Verduno Pelaverga, Colli Bolognesi Merlot or Carignano del Sulcis Rosso.

PORK CHOPS WITH APPLES

Serves 4

4 pork chops, 2 Reinette or similar crisp, sweet apples, 4 oz/100 g of leaf spinach, 2 tbsp of Calvados, 4½ tbsp/70 ml of cream, the juice of 2 lemons,
½ tbsp of mild mustard, 2 tbsp of demi-glace (rich brown sauce with Madeira or sherry), 4½ tbsp/50 g of clarified butter, 3 bay leaves

Method: 1. Peel the apples and cut them into 12 nicely shaped wedges, then brown them all round in a frying pan with a little butter, the juice of half a lemon and a tablespoonful of Calvados.
2. Blanch the spinach in boiling salted water, drain, then squeeze thoroughly in a cloth. Arrange a bed of spinach on each plate.
3. In a very hot sauté pan, fry the chops in a little clarified butter, seasoned with bay leaves. Pour off the fat from the pan, then deglaze with lemon juice and Calvados. When the liquid has evaporated, add the mustard, demi-glace and cream, and reduce by ⅔ to create a creamy sauce.
4. Strain the sauce, then pour over the chops. Serve garnished with apple wedges.

Wines: The predominantly sweet taste of the meat, apples and cream is perfectly balanced by the sharpness of the mustard and lemon juice. Serve with a strong, young, red wine with a well-developed fruity, floral bouquet, dry but still smooth, full-bodied and pleasantly acidic with very little tannin, such as Alto Adige Santa Maddalena, Colli del Trasimeno Rosso or Lacrima Christi Rosso.

LOIN OF PORK WITH HERBS AND MIXED BEANS FLAVORED WITH TARRAGON

Serves 10

4½ lb/2 kg of loin of pork, 1 head of fresh garlic, rosemary, sage, thyme, marjoram, extra virgin olive oil, 2 small onions, chopped, ¾ cup/150 g of cannellini beans, ¾ cup/150 g of borlotti beans, ½ cup/100 g black beans (soaked separately overnight in cold water), a bunch of tarragon, 7 tbsp/100 ml of white wine, ¾ cup/200 ml of vinegar, 3½ tbsp of sugar, ⅔ cup/150 ml of stock, the yolks of 5 eggs, 1 cup/100 g of grated Parmesan cheese, salt.

Method: 1. Boil the beans separately in salted water for as long as necessary (between 30 minutes and 1 hour, depending on the type).
2. Boil the tarragon, vinegar, sugar, and white wine in a saucepan for 20 minutes and set aside.
3. Bind the pork loin with kitchen twine. Chop the herbs and mix them with salt and rub the mixture all over the surface of the meat. Then place the joint in a lightly oiled roasting pan, add the head of garlic, unpeeled, and the chopped onion. Roast at 350°F/180°C for at least 30 minutes, adding wine or water from time to time, to avoid burning.
4. Bring the stock to the boil in a saucepan, take the pan off the heat and add the egg yolks, beating vigorously. Add the grated Parmesan cheese and return to the heat for three minutes. Add 5 oz/150 g of the beans and purée the mixture in a blender and then pass it through a fine sieve. Keep warm without allowing the sauce to return to the boil.

5. Brown a clove of garlic in a little oil, add the other drained beans, then the tarragon sauce. Reheat without boiling, and keep warm.
6. Remove the meat from the oven and leave to rest in the cooling oven for 20 minutes. Carve into slices and arrange the meat on a bed of beans and bean sauce. Pour the pan juices over the meat and serve.

Wines: The judicious choice of ingredients in this recipe provides a delightful balance of flavors, with the meat and beans providing sweetness, combined with the rich, aromatic tastes of the other ingredients. Serve with a mature, well-structured, strong red wine, fairly tannic, smooth and crisp with a bouquet of herbs, fruit and flowers, such as Piave Raboso, Sangiovese di Romagna Superiore or Montefalco Rosso.

COTECHINO IN VELOUTÉ SAUCE WITH WHITE TRUFFLES

Serves 4

1 very good-quality cotechino sausage, 1½ qt/1.5 l of skimmed chicken stock, 1 cup/100 g of vialone nano or other risotto rice,
1 cup/100 g of grated Parmesan cheese, 3½ tbsp/50 g of butter, 2 tbsp of extra virgin olive oil,
salt and pepper, truffles as required

Method: 1. Boil the *cotechino* in the usual way.
2. Meanwhile prepare the sauce. Pour the stock into a saucepan and bring to a boil. Add the rice, and stir briefly. For this recipe the rice should be well cooked.
3. Pour the stock and rice into a blender and grind to a thick purée. Turn the blender down to minimum speed, add the cheese, then the oil and, finally, the butter, cut into small pieces. As soon as the butter has been mixed in, transfer the mixture to a bowl and keep warm in a double boiler. Season lightly with salt and pepper, bearing in mind that the *cotechino* is already highly seasoned.
4. To serve, cut the *cotechino* into 2 in/5 cm slices, arrange in the center of a very hot plate, cover with velouté sauce and serve immediately with grated truffles, the more the better.

Note: This is a very simple recipe but to enjoy it at its best only use the finest-quality ingredients.

Wines: This is a perfect combination of traditional ingredients with innovative preparation and presentation, a simple and tasty dish with an interesting blend of flavors. To complement it, choose a delicate, young, red wine, not too strong, crisp with a hint of tannin and an intense bouquet of fruit and flowers, such as Valle d'Aosta Donnas, Chianti Colli Senesi or Leverano Rosso.

EZIO SANTIN

PORK CHOPS IN SWEET AND SOUR SAUCE

Serves 4

12 pork rib chops taken from the largest part of the rib, 4 tbsp of thyme honey, 3 tbsp of red wine vinegar, 1 tsp of balsamic vinegar,
1 tsp of ground mixed spices (ginger, black pepper, cardamom and star anise), 2 tbsp of olive oil, ⅔ cup/200 ml of stock made with pork bones,
salt and seasonings

Method: 1. Caramelize the honey in a small frying pan with the wine vinegar, but do not allow it to burn. Add a ladleful or two of stock and finally the balsamic vinegar and spices and simmer to produce a light syrup.
2. Brush the chops with the syrup. Leave the chops to marinate in the least cold part of the refrigerator overnight.
3. The following day, preheat the oven to 400°F/200°C. Heat the olive oil in a roasting pan and brown the chops all round over a high heat, season with salt, then cook in the oven for about 1 hour. Add stock from time to time, and baste the chops with the pan juices.

4. When the chops are cooked, pour off the fat in the pan, deglaze with the remaining stock, then reduce the sauce to a syrupy consistency and check the seasoning. Serve on very hot plates with the sauce and, if you like, a celery and turnip purée.

Wines: To accompany this tasty blend of savory, sweet and sour flavors, choose a very smooth, well-structured red wine, crisp and robust with just a touch of tannin. Try Valtellina Superiore Valgella, Morellino di Scansano or Salentino Rosso.

PORK FRY

Serves 4

8 tbsp of butter, 7 oz/200 g of pork fillet, half a lemon, 7 oz/200 g of pig's liver, 1 oz/30 g of pig's caul fat, 1 glass dry white wine,
1 sprig of rosemary, 1 onion, 1 qt/1 l of polenta, cooked, nutmeg, cinnamon, salt and pepper

Method: 1. Chop the onion and caul fat and brown in butter with the rosemary. Cut the liver into small slices, add to the pan and cook over a moderate heat for about 20 minutes.
2. Add the white wine and season with salt, pepper and lemon juice.
3. Slice the pork fillet, add to the pan and cook for a further 10-12 minutes. Serve hot with toasted polenta.

Note: Martinmas, or St Martin's Day, falls on November 11, as the countryside sleeps and stacks of grain add splashes of color to the chilly fields. The arrival of the fog and cold of winter also marks the beginning of the pig-slaughtering season. The ritual is faithfully repeated in every household and, after a hard day's work, everybody gathers round the table. All the friends, relations and neighbors who helped with the many tasks of skinning, cutting, chopping, grinding and curing the meat, and of

mixing, stuffing and binding the sausages, sit down together for a celebratory feast. Custom demands that they eat rice and cabbage with ground salami and boiled bones. Black pudding, good Lambrusco and traditional accordion music enliven the evening. The following day, it starts all over again in another house. Pork fry is served at lunchtime.

Wines: This succulent traditional dish from the Mantua region is spicy and aromatic, with a slightly bitter aftertaste. It should be accompanied by a robust but smooth young red wine, with a rich, fruity bouquet and just a little tannin, such as Barbera d'Asti, San Colombano or Pentro di Iserna Rosso.

NADIA SANTINI

RICE WITH MINCED PORK AND CABBAGE

Serves 4

For the minced pork: 9 oz/250 g of fresh pancetta, 9 oz/250 g of fresh pork shoulder, 1 clove of garlic, crushed, 2 tbsp of red wine, 2 level tsp of salt.
For the risotto: 1 oz/30 g of chopped meat, 1 oz/30 g of chopped onion, 1 oz/30 g of chopped celery, 4 oz/100 g of finely chopped cabbage,
3 cups/350 g of vialone nano or other risotto rice, 1 ladleful of stock, 2 cups of water, 2 tbsp of butter.

Method:
1. Mince the fresh pork and mix it with the salt, wine and garlic. In a saucepan, brown the onion, celery and cabbage in butter. Add the minced pork and the chopped meat and simmer for 7-8 minutes, then add the stock.
2. Add the rice and boiling water, stirring constantly. Cook for 12-13 minutes and serve immediately. By tradition, the rice is served in a broth, but it is also good served with butter.

Note: This was a traditional winter dish enjoyed in many parts of the Po Valley. It was customary to offer it to the people who assisted the *norcino* (pork butcher) to chop the freshly slaughtered pig and make sausages.

Wines: The smoky flavor of the rice and the spicy taste of the other ingredients are best complemented by a crisp, fairly strong, young red wine with a fruity, floral bouquet, and not too much tannin. Choose a Grignolino del Monferrato, Friuli-Isonzo Refosco or Cilento Rosso.

ROMANO TAMANI

PORK FILLET WITH ONIONS, APPLES AND TOMATO SAUCE

Serves 4

12 small pieces of pork fillet weighing about 1 lb/450 g altogether, 4 medium red onions, ½ cup/100 g of passata (sieved tomatoes), 1 large, crisp apple, 8 peppercorns, 7 tbsp/100 g of butter, salt to taste

Method: 1. Peel the onions and cut them into fairly thick rings. Melt the butter in a large frying pan, add the onion rings and cook over a low heat for 15 minutes. Add the salt, peppercorns, and passata, and cook for 5 minutes.

2. Peel and core the apple and cut it into 12 wedges. Add the meat and apples to the pan and cook for 5 minutes.

3. Arrange three fillets on each plate with three apple wedges and a few onion rings. Sieve the rest of the sauce and pour over the meat.

Wines: Here we have a pleasing contrast between the savory pork, the sweet aromatic flavors of apple and onion, and the slight acidity of the tomatoes. This dish calls for a young, dry, red wine, delicate and pleasantly acidic, with only a hint of tannin and a fruity, floral bouquet, such as Oltrepò Pavese Cabernet Sauvignon, Taburno Rosso or Orta Nova Rosso.

ROMANO TAMANI

STEWED PORK AND PEAS

Serves 4

5 oz/150 g of lean loin of pork, 5 oz/150 g of leg of pork, chopped sage, rosemary, garlic and oregano, 7 oz/200 g of fresh peas, shelled, 4 ripe tomatoes, 1 glass of dry white wine, 7 tbsp/100 g of butter, ¼ cup/50 ml of oil, 1 onion, finely chopped, salt and pepper to taste, flour as required, 4 cabbage leaves, blanched

Method: 1. Cut the meat into cubes of equal size. Melt the butter in a frying pan with the oil and fry the onion until golden. Coat the meat in flour and sauté for a few minutes with the onion. Add the white wine and simmer until evaporated.

2. Peel and dice the tomatoes and add to the pan with the peas, chopped herbs, salt and pepper. Cook, covered, for about 2 hours over a moderate heat, to create a thick stew.

3. Arrange a blanched cabbage leaf on each plate and top with very hot stew.

Wines: In this recipe, the sweetness of the pork and peas combines with the aromatic flavors of the other ingredients. Serve with a young, dry, perhaps lightly sparkling, red wine, medium bodied, smooth and pleasantly acidic, with a rich bouquet of fruit and flowers, such as Oltrepò Pavese Buttafuoco, Colli Piacentini Bonarda or Costa d'Amalfi Tramini Rosso.

GRILLED FILLET OF PORK IN RED WINE SAUCE WITH POPPY SEED GNOCCHI AND SAUTÉED BLACK OLIVES

Serves 4

2 pork fillets weighing about 1½ lb/600 g altogether, 4 oz/100 g of black olives, 2 tbsp of butter, a generous pound/500 g of potatoes,
1½ cups/170 g of flour, ½ cup/50 g of poppy seeds, the yolks of 2 eggs, extra virgin olive oil
For the sauce: ½ lb/240 g of onions, 2 bottles of full-bodied red wine, 3½ tbsp of red wine vinegar, 6 shallots, 4 sprigs of thyme, 1 cup of port,
the juice of 1 lemon, 2 stock cubes, 1 cup/250 ml of meat stock, ½ cup/100 g of caramelized sugar,
2-3 tsp of potato starch or cornstarch, salt

Method: 1. Coarsely chop the onions, place them in a saucepan with the wine and boil until reduced by half. Add all the other sauce ingredients, except the potato starch, and boil for a further 10 minutes. Pass the sauce through a fine sieve, then return it to the saucepan and bind with the potato starch (dissolved in a little cold water).
2. Boil and peel the potatoes. Blend them with the flour and egg yolks to make a smooth dough then shape the dough into tiny gnocchi. Boil the gnocchi in salted water and, when they float to the surface, remove with a slotted spoon and sauté them in butter with the poppy seeds.
3. Remove the pits from the olives and sauté them in a frying pan with olive oil.
4. Grill the pork fillets and carve them into ½ in/1cm slices.
5. Arrange the meat on a serving dish, attractively garnished with gnocchi and olives. Add the sauce and serve.

Wines: This typically Mediterranean dish with its blend of savory, aromatic flavors and slightly bitter aftertaste should be accompanied by a mature, well-structured red wine, strong, smooth and discreetly tannic, with a bouquet of red fruits and spices. Try Nebbiolo d'Alba, Carso Terrano or Rosso Cònero.

SPICED LOIN OF PORK WITH HONEY AND ORANGE SAUCE AND ONION COMPÔTE

Serves 4

1½ lb/600 g of loin of pork, boned and cut into four, 4 slices of pancetta or streaky bacon, 2 apples, ¼ cup/50 g sugar, vegetable oil,
equal quantities of crushed pepper, mace, cinnamon and juniper berries
For the onion compôte: a generous pound/500 g of white onions, 7 tbsp of vinegar, 3 cups/750 ml of red wine, vegetable oil
For the sauce: 3-4 tbsp of acacia honey, 3½ tbsp/50 g of sugar, the juice of 2 oranges, 6 tbsp/100 ml of vinegar, 1 qt/1 l of brown veal stock

Method: 1. To make the sauce, caramelize the sugar and honey, then add the orange juice and vinegar. Add the veal stock and reduce to a syrupy consistency. Reserve the sauce.

2. To make the onion compôte, peel and wash the onions and cut them into julienne strips. Fry gently in a little oil, without allowing them to brown. Add the vinegar and enough water to cover. Boil until all the liquid had evaporated, then add the wine and continue to cook until completely absorbed. Reserve the compote.

3. Peel and core the apples, cut them into wedges, then use a melon baller to cut them into balls the size of an olive. Boil in water with sugar to taste.

4. Season the meat with salt and pepper, rub with the spices, then wrap each portion in a slice of pancetta, securing it with kitchen twine. Heat the vegetable oil in a frying pan and brown the meat all round.

5. Pour the sauce into the center of each plate. Untie the meat and arrange it on top of the sauce. Garnish with the onion compôte (if you like, shaped into croquettes using two spoons) and the apple balls. Sprinkle with spices and serve immediately.

Wines: To complement the delicious combination of succulent, spiced meat and the sweet and sour flavors of the other ingredients, serve a fairly mature, well-structured and distinctly alcoholic red wine. Choose something very smooth, with minimal tannin and a rich bouquet of ripe fruits, red flowers and spices, such as Valle d'Aosta Enfer d'Arvier, Colli Bolognese Cabernet Sauvignon or Sant'Agata dei Goti Rosso.

| UMBERTO VEZZOLI |

SMOKED PORK AND CABBAGE SALAD

Serves 4

7 oz/200 g of smoked loin of pork, 4 oz/100 g of cabbage, 3½ tbsp/50 ml of Barolo or other good red wine vinegar,
½ cup/100 g of tomatoes, peeled, deseeded, and diced, 1 cup/50 g of parsley, ¾ oz/20 g of anchovies preserved in salt,
2 tsp of extra virgin olive oil, salt and pepper to taste

Method: 1. Trim any fat from the meat and boil in unsalted water for about 30 minutes. Allow the meat to cool in its own stock.
2. Shred the cabbage very finely.
3. Make a dressing with the vinegar, olive oil, finely chopped anchovies and a little salt and pepper, and keep it in the refrigerator until required.
4. Arrange a bed of shredded cabbage in the center of each plate and lay the thinly sliced meat on top. Garnish with diced tomatoes and parsley leaves, pour over the dressing, and serve at room temperature with walnut bread.

Wines: This tempting example of contemporary cuisine, is a delightful combination of succulent meat and crunchy cabbage, harmoniously balanced by the distinctive flavors of the dressing. Serve it with a very delicate, dry white wine, fairly acidic, strong and full-bodied, with an aromatic bouquet, such as De Vite, Carso Malvasia or Regaleali Bianco.

UMBERTO VEZZOLI

BOILED PIG'S CHEEK WITH GARBANZA BEANS

Serves 4

11 oz/300 g of fresh pig's cheek, 1 cup/200 g of dried garbanza beans, 4 oz/100 g of onion, 4 oz/100 g of celery, 2 sprigs of sage, 2 sprigs of rosemary,
1¼ cups/300 ml of stock, 3 ripe tomatoes, chopped, 4 oz/100 g of carrots, 7 tbsp/100 g of butter, extra virgin olive oil,
salt and pepper to taste

Method: 1. Soak the garbanza beans overnight in cold water.
2. Singe the pig's cheek over a flame to remove all traces of bristle from the rind, then boil in water with some of the celery, carrots, onion, and sage over a low heat for 2 hours. Skim the scum from the surface of the water and, once no more scum rises, add salt.
3. Chop the remaining vegetables and sauté them in a frying pan with oil and butter. Drain the garbanza beans thoroughly and stir them into the vegetables with the chopped tomatoes. Add the stock and rosemary and simmer for about 2 hours.
4. When all the ingredients are cooked, arrange the garbanza beans on a serving dish with the sliced meat on top. Serve very hot, drizzled with extra virgin olive oil and sprinkled with coarse salt and coarsely ground black pepper.

Wines: The succulence and sweetness of the pig's cheek and garbanzo beans is offset by the delicate flavor of herbs and the subtle richness of olive oil. To accompany this dish, serve a strong, young, dry red wine, very crisp and slightly tannic, with an intense bouquet of fruit and flowers, such as Ruché di Castagnole Monferrato, Friuli-Aquileia Refosco dal Peduncolo Rosso or Cannonau di Sorres.

PIGS' KIDNEY AND PORCINI CASSEROLE WITH GRILLED POLENTA

Serves 4

1½ lb/600 g of very fresh pigs' kidneys, 11 oz/300 g of fresh porcini mushrooms, 2 shallots, chopped, 7 tbsp of brown veal stock, 7 tbsp of sunflower oil, 1 cup/50 g of chopped parsley, 7 tbsp/100 ml of Pinot Bianco wine, salt and pepper

Method: 1. Clean and trim the mushrooms. Thinly slice the kidneys and mushrooms.
2. Brown the shallots in a long-handled frying pan, then add the kidneys and mushrooms and sauté for about 2 minutes.
3. Add the wine and cook until evaporated.
4. Add the veal stock and parsley, then quickly reduce, and serve with slices of fine polenta, grilled.

Note: The secret of success with this recipe is to add the ingredients as quickly as possible, since kidneys can be very tough if overcooked.

Wines: In this dish, the porcini mushrooms and shallots add delicious, subtle flavor to the kidneys. Serve with a well-structured, crisp, dry white wine, very smooth and strong, with a well-developed and intense bouquet of fruit and flowers. Try Torre de Giano riserva, Ischia Biancolella or Corvo Colomba Platino.

PORK FILLET AU GRATIN WITH AROMATIC HERBS

Serves 4

*1 pork fillet weighing 1½-1¾ lb/700-800 g, 2½ cups/300 g of stale breadcrumbs, 2 eggs, 4 oz/100 g of thyme, rosemary, marjoram and sage,
1 cup/100 g of grated Montasio stravecchio or other piquant, crumbly cheese, 7 tbsp/100 ml of sunflower oil, 4 tbsp/50 ml of extra virgin olive oil*

Method: 1. Chop all the herbs. Add the breadcrumbs, cheese and a little olive oil.

2. Season the meat with salt and pepper, coat it with beaten egg, then cover it with the herb and breadcrumb mixture.

3. Preheat a frying pan with sunflower oil, then fry the meat for about 3 minutes until brown all over. Place the meat on a grid over a roasting pan and roast in the oven at 325°F/160°C for about 10 minutes.

4. Carve the meat into thin slices and serve with crisp cabbage prepared with smoked pig's cheek.

Wines: Herbs and mature cheese lend this dish strong, aromatic flavors to contrast with the sweet and succulent meat. To accompany it, choose a full-bodied, medium mature, red wine, pleasantly acidic with a touch of tannin and an intense bouquet of fruit and flowers, such as Lessona, Ornellaia or Contea di Sclafini Syrah Riserva.

We wish to thank all those who have helped in
the production of this book, especially:
Bibliotheca Internazionale di Gastronomia
Sorengo, Switzerland

Gastronomia Peck, Milan

Antica Macelleria Cecchini,
Panzano in Chianti

Dr. Paola Salvatori

Dr. Marta Lenzi

CHEFS AND RESTAURANTS WHO CONTRIBUTED TO THIS BOOK

Pinuccio Alia
La Locanda DI ALIA
contrada Jetticelle, 69
87012 Castrovillari (CS)

Roberto Andreoni
Ristorante VIA DEL BORGO
viale della Libertà 136
20049 Concorezzo (MI)

Darko Bank, Elvio Muha,
Paolo Polla
Ristoraante Buffet DA PEPI
via Cassa di Risparmio, 3
34100 Trieste

Bruno Barbieri
Locanda SOLAROLA
via Santa Croce, 5
40023 Castel Guelfo (BO)

Sergio Bartolucci
Ristorante EUROSSOLA
piazza Stazione, 36
28037 Domodossola (VB)

Karl Baumgartner
Restaurant SCHÖNEK
via Castel Schönek, 11
loc. Molini
39030 Falzes (Bz)

Walter Bianconi
Ristorante TIVOLI
via Lacedel, 34
32043 Cortina d'Ampezzo (BL)

Luca Bolfo e Mario Oriani
Ristorante VECCHIO MULINO
via al Monumento 5
27012 Certosa di Pavia (PV)

Gianni Bolzoni
Trattoria de FULMINE
26017 Trescore Cremasco (CR)

Aurelio Bonardi
Il SALUMAIO di
Montenapoleonie
via Monte Napoleone 12
20121 Milano

Luihif Bortolini
Ristorante DA GIGETTO
via A. De Gasperi, 431050
Miane (Tv)

Anne e Lucia Botte
Ristorante CERESOLE
via Ceresole, 4
26100 Cremona

Sauro Brunicardi
Ristorante LA MORA
via Sesto di Moriano, 1748
55029 Ponte a Moriano (LU)

Vera Caffini
Ristorante AQUILA NIGRA
vicolo Bonacolsi, 4
46100 Mantova

Sergio Carboni
Albergo Ristorante ITALIA
via Garibaldi, 1
26038 Torre de Picenardi (CR)

Marco Cavallucci
Ristorante LA FRASCA
viale Matteotti, 34
47011 Castrocaro Terme (FO)

Dario Cecchini
Antica Macelleria CECCHINI
via 20 Luglio, 11
50020 Panzano (Fi)

Filippo Chiappini Dattilo
Antica Osteria DEL TEATRO
via Verdi, 16
29100 Piacenza

Georges Cogny
Locanda CANTONIERA
S.S. 654 - 29023 Farini (PC)
tel. 0523/919113

Igles Corelli
Locanda della TAMERICE
via Argine Mezzano, 2
44020 Ostellato (FE)

Enzo De Prà
Ristorante Albergo DOLADA
via Dolada, 21
32010 Plois in Pieve d'Alpago (BL)

Enrico Derflinger
Hotel EDEN
via Ludovisi, 49
00187 Roma

Mario Di Remigio
Ristorante SYMPOSIUM
QUATTRO STAGIONI
via Cartoceto, 38
61040 Cartoceto (PS)

Corrado Fasolato
Ristorante LA SIRIOLA
loc. Armentarola in Pre de Vi, 127
39030 San Cassiano (BZ)

Massimo Ferrari
Ristorante AL BERSAGLIERE
via Goitese, 260
46044 Goito (MN)

Walter Ferretto
Ristorante IL CASCINALE
NUOVO
Statale Asti-Alba, 15
14057 Isola d'Asti (AT)

Alfonso Iaccarino
Ristorante DON ALFONSO 1890
corso S. Agata, 11
80064 S. Agata sui due Golfi (NA)

Ernst Knam
L'ANTICA ARTE DEL DOLCE
via A. Anfossi, 10
20135 Milano

Angelo Lancellotti
Ristorante LANCELLOTTI
via A. grandi, 120
41019 Soliera (MO)

Valentino Marcattilii
Ristorante SAN DOMENICO
via G. Sacchi, 1
40026 Imola (BO)

Sergio Mei
Hotel FOUR SEASONS
via Gesù, 8
20121 Milano

Fabio Momolo
Hotel SPLENDID
Viale delle Terme
35030 Galzignano terme (PD)

Aimo and Nadia Moroni
Ristorante AIMO E NADIA
via Montecuccoli, 6
20147 Milano

Arneo and Dario Nizzoli
Ristorante NIZZOLI
via Garibaldi, 18
46030 Villastrada di Dosolo (MN)

Davide Oldani
Ristorante GIANNINO
via A. Sciesa, 8
20135 Milano

Fabio Picchi
Ristorante CIBREO
via dei Macci, 118 r
50122 Firenze

Claudio Prandi
Hotel Ristorante IL GRISO
via Provinciale, 51
22040 Malgrate (LC)

Adriano Presbitero
Hotel Ristorante PANORAMICA
via San Rocco, 7
10010 Loranzé (TO)

Romano Resen
Hotel PRINCIPE DI SAVOIA
piazza della Repubblica, 32
20124 Milano

Elia Rizzo
Ristorante IL DESCO
via San Sebastiano, 7
37121 Verona

Andrea Rossi
Ristorante PECK
via Spadari, 9
20121 Milano

Romano Rossi
Trattoria IL TESTAMENTO
DEL PORCO
via O. Putinati, 24
44100 Ferrara

Claudio Sadler
Ristorante SADLER
via Troilo, 14
20136 Milano

Ezio Santin
ANTICA OSTERIA DEL
PONTE
piazza G. Negri, 9
20080 Cassinetta di Lugagnano
(MI)

Nadia Santini
Ristorante DAL PESCATORE
46013 Runate di Canneto
sull'Oglio (MN)

Romano Tamani
Ristorante AMBASCIATA
via Mattiri di Belfiore, 33
46026 Quistello (MN)

Paolo Teverini
Ristorante TOSCO-
ROMAGNOLO
piazza Dante, 2
47021 Bagno di Romagna (FO)

Umberto Vezzoli
Hotel PALACE,
CASANOVA GRILL
piazza della Repubblica, 20
20124 Milano

Luigi Zago
Locanda alle OFFICINE
via Nazionale, 46/48
33042 Buttrio (UD)

MAIN COURSES

DESSERTS

un spagheto, è come un sacchetto ...
chiuso, è fauvi fumo di ginepro, d'a...
fuoco, accio di colino, è come stan...
vali a 4., o 5. serrate, è vimethile...
sodi, si conserverlo methili sotto ...
nell'acqua si lo spazio di 3., o 4. ...
lasciali stendere in d. acqua, ...
che si mettii dentro, è non essendo asciug...
bra di carne, perchè di sala come ...